THE SKY BEYOND

"The stories in this book cover the first thirty-five years of my flying life. Those years were lived through two world wars and most of the pioneering age of transoceanic flight which has led to the international air services of today.

"It has been said that I am lucky to have survived to write this book; and that is true: but luck alone is not enough; for the Air, like the Sea, is an uncompromising though rewarding element. It demands a certain standard of its close associates. The price of falling below that standard can be high. The urge always to seek and to know is strong within us and is part of the challenge and the inspiration that keeps us flying."

From the Foreword by
Sir Gordon Taylor

THE AVIATOR'S BOOKSHELF
THE CLASSICS OF FLYING

THE SKY BEYOND

SIR GORDON TAYLOR

BANTAM BOOKS
Toronto • New York • London • Sydney

THE SKY BEYOND
*A Bantam Book / published by arrangement with
Houghton Mifflin Company*

*PRINTING HISTORY
Bantam edition / November 1983*

ABOUT THE COVER ARTIST

WILSON HURLEY is a well-known landscape artist. His aviation and space paintings hang in the Air Force Museum, the Air Force Academy, and at NASA headquarters. The Greenwich Workshop, 30 Lindeman Drive, Trumbull, CT 06611, publishes Wilson Hurley's aviation and space paintings as well as his landscapes.

ISBN 0-553-23949-X

*Published simultaneously in the United States
and Canada*

Bantam Books are published by Bantam Books, Inc. Its trademark, consisting of the words "Bantam Books" and the portrayal of a rooster, is Registered in U.S. Patent and Trademark Office and in other countries. Marca Registrada. Bantam Books, Inc., 666 Fifth Avenue, New York, New York 10103.

PRINTED IN THE UNITED STATES OF AMERICA

H   0 9 8 7 6 5 4 3 2 1

Foreword

THE STORIES in this book cover the first thirty-five years of my flying life. Those years were lived through two world wars and most of the pioneering age of transocean flight which has led to the international air services of today.

It was inevitable that risks had to be weighed in the balance against the purpose of our ventures, and often accepted as a worthwhile condition of them: so I feel some responsibility to explain that passengers traveling on today's airlines are not exposed to the situations about which they may read in this book. But part of our contribution to the very high safety standard of the airlines is the experience gained in giving a lead to the services which have followed upon our tracks.

It has been said that I am lucky to have survived to write this book; and that is true: but luck alone is not enough; for the Air, like the Sea, is an uncompromising though rewarding element. It demands a certain standard of its close associates. The price of falling below that standard can be high. The urge always to seek and to know is strong within us and is part of the challenge and the inspiration that keeps us flying.

There is always the incentive to reach out into the sky beyond, and that perhaps has dominated the period of my flying life which is covered in this book; for far in the depths of night above the ocean, particularly where no aircraft has flown before, there is a vast tranquillity filled with vivid life, a calm and certain knowledge of the most natural immortality of which one is part with all creation. To fly in the presence of this knowledge is a humbling but deeply significant experience which erases all trivialities, and one returns to Earth calmly, equipped again, if only temporarily, with a sense of spiritual fulfillment and of physical well-being.

P. G. TAYLOR

Contents

for weather, were gone forever. The A.N.A. Fokkers went

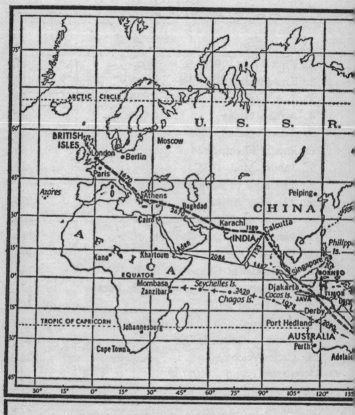

Trans-Tasman experimental flights in SOUTHERN CROSS: 1933
1934, 1935
ALTAIR, first west to east crossing of Pacific Ocean, 1934
Solo record flights in PERCIVAL GULL, Java to Sydney;
June 1934, and May 1937
First crossing and survey flight of Indian Ocean, June 1939

Distances in nautical miles

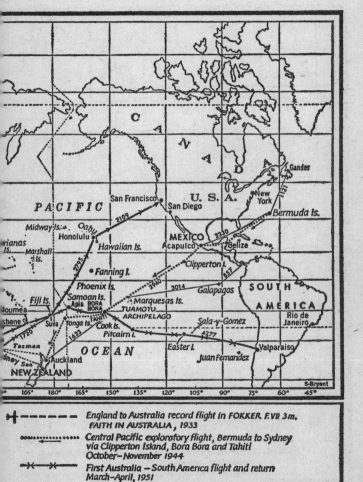

C A N A D A

PACIFIC

San Francisco
San Diego
U.S.A.
New York
Gander

Midway Is.
Oahu
Honolulu
Hawaiian Is.
2109
MEXICO
Acapulco
2330
Belize
Bermuda Is.

arianas Is.
Marshall Is.
2705
• Fanning I.
Clipperton I.
657

Phoenix Is.
2940
3014
Galapagos
SOUTH

Fiji Is.
Samoan Is.
Apla BORA BORA
TAHITI
Marquesas Is.
TUAMOTU ARCHIPELAGO
AMERICA
Rio de Janeiro

Noumea
Suva
Tonga Is.
Cook Is.
Pitcairn I.
Sala-y-Gomez

sbane
1720
1623
Easter I.
4377
Valparaiso

Tasman
Juan Fernandez

ney Sea
Auckland
OCEAN
S·Bryant

NEW ZEALAND

165°   180°   165°   150°   135°   120°   105°   90°   75°   60°   45°

+ — — — —   England to Australia record flight in FOKKER F.VII 3m.
               FAITH IN AUSTRALIA, 1933

•••••••••••••   Central Pacific exploratory flight, Bermuda to Sydney
               via Clipperton Island, Bora Bora and Tahiti
               October–November 1944

× — × — ×   First Australia – South America flight and return
               March–April, 1951

THE SKY BEYOND

Chapter 1
First Flight, 1916

EARLY in the morning of August 5, 1914,* groups of boys crowded round the newspaper stands in the school library. The dramatic news was there in the staring headlines. England had declared war on Germany.

I was one of those boys, with still another year at school before I would be old enough to go. Our picture of war was a glorious affair in those times—a terrific adventure of drums and flags and soldiers charging into battle, routing the enemy. It was by pure chance that I happened to join the Royal Flying Corps, and to become a fighter pilot on the western front in France.

I was reading the *Strand Magazine,* coming home in the train for the holidays. There was a story of British aircraft bombing the Zeppelin hangars at Cuxhafen, with illustrations of the Avros diving on the great sheds which housed the German airships. It all seemed exciting and adventurous. This, I thought, was obviously the way to go to war.

So, in 1916, after discovering that the recently formed

---

* Because of time difference between England and Australia the English declaration of war on August 4 was only published in Australia on the morning of August 5.

Australian Flying Corps was not at that time enlisting people
for training as pilots, I went to England and joined the Royal
Flying Corps. I did not realize that the ease with which I got
into the R.F.C. was partly due to the very heavy casualties
being inflicted by new German aircraft upon the crews of the
obsolete types with which the R.F.C. was very largely
equipped.

I passed through a course of ground instruction at Reading, was posted to Netheravon elementary flying training
school on Salisbury Plain, and the adventure of the air was
before me. After more lectures on aircraft and some frustrat-
ing days of waiting around the hangars for flying training to
begin, I was suddenly confronted by a rather morose instructor who, with few words, launched me abruptly into my first
experience of flight.

"You Taylor?"

"Yes, sir."

"Had any instruction yet?"

"No, I haven't, sir."

"All right, I'll give you some landings."

With these brief words he walked off towards a Maurice
Farman standing on the tarmac and I followed him in a
somewhat confused state of mind. This wasn't exactly what I
had expected for the beginning of my flying training. I had
pictured some sort of orderly approach to the thing: a talk
with the instructor before my first flight, and some idea of
how to control the airplane. Instead, I climbed up after him
into the rear seat of a thing about the size and shape of a
bathtub, slung between the wings of a biplane held together
by many wires and struts and having the engine with a
pusher airscrew located immediately behind the place where I
was sitting. Though I was charged with enthusiasm for the
great adventure which was about to begin, I couldn't help
noting that the engine was conveniently located to come right
through the back of my neck in the event of a crash dislodg-
ing it from its mountings.

But there was little time for speculation. A mechanic
swung the propeller, the pilot ran up the engine, waved away
the chocks, and we were away out onto the grassy ridge of
the airfield. My instructor said nothing. I sat in the back con-

fronted with what looked like the handle of a large pair of scissors on the top end of a control column, and a rudder bar on which I didn't know whether or not to put my feet. Having received no instructions I thought it best not to touch anything; so I kept my hands and feet off the controls.

We swung round into wind, the sound of the engine rose to a roar behind me and the machine started to move off over the surface. In what seemed to me to be a very short distance the rattling of the undercarriage on the grassy surface subsided and stopped and I realized that we were flying. Rather than having any sensation of climbing I felt that the ground was sinking away from the airplane and a new landscape invisible from below was spreading out around us. I waited in excited anticipation for some words from the instructor; but none came.

Instead, the airplane turned up on a wing, circled the airfield, and faced up in the direction in which we had taken off. Quite suddenly the roar of the engine ceased, the nose went down, and we were gliding steeply for the earth. I sat back keeping clear of everything and wondered whether this was a normal approach to the ground or something had gone wrong and we were just going to crash. The earth rushed up at us, the airplane flattened quite smoothly out of its steep glide, and in a few moments I again felt the rattling of the wheels on the grass.

Now, I thought, we shall stop and I shall be told how to do it. But no, again. Before the machine had come to rest the engine came on with a roar and we were away again, and into the air. We went into the same close circuit of the airdrome, flew downwind, and round, to line up for the approach to land. The engine was throttled back, the nose went down, and a shout came from the instructor in the front seat, "You land it!"

The shock of this remark caused a physical reaction almost before my mind had grasped the enormity of its meaning. My hands shot out to the scissors handles in front of me and my feet to the rudder bar on the floor. Instinctively I imagined that unless I took the controls instantly the airplane might fall out of control to the ground. I knew nothing. Nothing beyond the absurdity of the demand that I should land the airplane. Then the first spark of hostility was ignited in the panic that threatened to engulf me in this ridiculous situation. This man was in the airplane with me. Even if he was mad enough to expect me to land it with no instruction he would probably be sane enough to stop me from crashing it.

I tried to hold the controls steady so that the machine continued on down in the glide he had set as he handed over to me. I daren't in fact move them much because I had no idea of the effect of movement of controls. Somehow I managed to keep the old Rumpety going on down. There wasn't much height to lose anyway: only about two hundred feet. Very soon the earth was visibly rushing up to meet us and I realized that something would have to be done before we dived into the ground. I drew back on the control column, much over-controlling of course, and the airplane began to swoop up again. The engine came on, the controls were snatched out of my hands and a savage shout came back to me from the instructor, "Bloody awful!"

Again we climbed away, went sweeping round the circuit and in for another landing. I tried mentally to anticipate the next demand, but none came and I just sat there, out of the picture, feeling utterly confused and ineffectual. The grass was coming up again and another landing was upon us. The machine began to flatten out, swept low over the surface and the now familiar clatter of the wheels told me we were on the ground once more. Visibly acknowledging my existence for the first time, the instructor turned slightly and announced in a more conciliatory tone, "That's better."

"What's better?" I thought. Then the impact of his meaning hit me, and with it a surge of fear about this whole mad act. I hadn't touched the controls. I still have no idea how the airplane was landed that day. I wanted to get out of it now: out, onto the ground, away from this man, and his flying machine.

But before I could utter a word in protest the engine at the back of my neck shattered all coherent thought with another blast of urgent power, the propeller thrust the machine forward, and we were away again. I mentally clung to the sides of the airplane, torn now between a conviction that I must somehow get out of this the next time we touched the ground, and a shocked impression of the consequences from such a flagrant breach of military discipline. Thoroughly shaken now by this conflict, and with the threat of disaster dominating all my reactions, I shirked my decision, and with the last shreds of morale which were left to me waited for the next move.

It surprised me. Instead of another mad circus performance round the airfield we climbed steadily and flew away over Salisbury Plain.

I was now given some brief instruction in how to fly the airplane straight and level, and found, with a returning sense of inspiration, that this was fairly easy. The instructor also seemed to have settled down and I felt for the first time some reasonable sort of human contact between us.

After about twenty minutes' practice in the most elementary control in level flight the airplane was again snatched out of my hands without comment, we went soaring round and back for the field. Obviously we were going to land again, and with that realization all the warning signals went up again. Was I to land it? Was he? What should I do? The machine banked into a turn and the engine stopped. We were already in the final glide. Too late to say or do anything. I resigned myself, and waited.

The instructor did the landing, and, to my very great relief, turned the airplane and taxied in for the hangars. On the tarmac he switched off the engine, climbed out of the nacelle, jumped down to the ground, and walked away.

So this was learning to fly.

I sat for a few moments, taking stock. Then I too got out of the airplane and went down to the quarters. I badly needed some thought and some advice about this whole affair. Having absorbed a strong traditional sense of military service and the sanctity of military discipline from recent experience at school where I was an officer in the Cadet Corps and senior prefect of the school, I had instinctive respect for authority and an inclination to accept and obey an order from a senior officer without question even in my mind. This instructor was of course senior in rank to me: a captain, against my recently acquired status in the Royal Flying Corps as a temporary second lieutenant on probation.

Beneath all this was a fundamental sense of injustice and resentment at so disillusioning and untidy an experience for my first flight in an airplane. I had come twelve thousand miles from Australia, with high hopes and inspiration to fly and fight in the service of the Royal Flying Corps. I knew I could sail a boat, ride a horse, run, swim, play football, and do most of the things I had been asked about as physical qualifications for service as a pilot in the R.F.C. I didn't believe I was such a fool as this instructor seemed to think. But I was confused and frustrated with conflicting thoughts and emotions after this wretched beginning to my career as a pilot.

That evening I had a good talk with "Anzac" Whiteman, a Rhodes scholar from Perth in Western Australia. Whiteman and I had gone through ground school together at Reading. We had teamed up as friends, and formed the idea of trying to go right on through to a squadron in France. I felt better after a talk with Anzac: decided to turn up for flying the next day without any approach to the instructor, and have the situation out with him in the airplane if things went the same way.

On the tarmac he was perfectly friendly, even cordial, as we went out to the machine. But as soon as we were in our seats and our belts fastened he started the engine and went off without explanation or any indication of what we were going to do.

In the air we did a few very gentle turns in which I had a somewhat indefinite part, not knowing really whether he, or I, was flying the airplane: a few minutes straight and level again; and then, with mixed emotions and a "this is it" feeling, I realized that landings were upon us again.

This time I was told to do the landing: a shout from the front, with no details. I had discovered that if you let the nose of this airplane go down beyond a certain speed you could not pull it out; also that if you didn't put it down far enough it would stall and crash. So I took the controls and really concentrated on the critical action of keeping the airplane in the correct glide, of easing her out, and touching the wheels down on the grass. I could feel the instructor also on the controls, creating an uncertainty in opposition to my movements; and in this way we went on down, both landing the airplane. In some intangible fashion the landing turned out to be quite a good one; but I had had enough. I was afraid and angry, and determined now to have a showdown with this man before another take-off.

He didn't give me a chance. To my utter astonishment he got up out of his seat, climbed over the side and down, to stand on the grass of the airfield. Then he looked up and called to me, "You can bloody well go solo now."

For the moment the shock froze me into a state of inaction. I couldn't believe he was serious. Then he started to walk away. This was it. I had to make a decision, for better or worse. I put my hand over the side of the nacelle, felt for the switch, and stopped the engine.

The instructor turned and came back, shouting at me, "What's the matter with you? Why did you stop the engine?"

Horror-stricken momentarily by the implications of my decision, I hit back at him. "I am not going solo."

I expected, and was quite prepared, to be instantly placed under arrest for refusing to obey an order. I knew I couldn't fly the airplane. I had never done a take-off, nor a landing without interference, and in the total of one hour and thirty minutes of this nightmare experience had briefly flown the machine straight and level alone. I felt hostile now; not horrified by fear of the consequences any more. I hadn't come all the way from my home in Australia to kill myself at the orders of this lunatic. But I was shaken and desperately unhappy about it. I waited for the blast I expected would be forthcoming. But this man was completely unpredictable. He called up to me, in an almost friendly voice, "What do you want to do?"

What *did* I want to do? A quick, instinctive thought shot out of my mind as an innocent statement of fact. "I want to fly with another instructor."

"Very well then; we'll taxi in and I'll see if I can arrange it for you." There was even a note of relief in his voice. Was I as bad as all that?

In at the hangar I was handed over to a long, genial instructor. His name was Prallé. Lieutenant Prallé. I see his name now, signed in my logbook. He wasted no time. "I believe you're having some trouble with your flying, Taylor?"

"Well, not exactly, sir. I just don't know how to fly, that's all."

"All right. Come on out and we'll start from scratch."

From that moment I knew I was on the way. Prallé gave me another two hours' concise and friendly instruction: take-offs, climb, straight and level flight, gentle and medium turns, glide with engine off, and landings: and a firm warning not to let the Rumpety's nose down below the angle of glide and speed he had shown me.

I went off solo keyed up to the realization of flight alone, where nobody could help me but myself, but confident; and in a few more days finished my time on Rumpetys and went on for advanced training at the Central Flying School, a few miles away at Upavon.

But Anzac Whiteman was killed; on his first solo he got the nose down and flew into the ground.

Some time afterwards I heard the story behind my first instructor. He had been a good pilot, with a fine record in France; but finally cracked up under the stress of flying an airplane of poor performance and ineffective armament against the new German fighters. Sent home for a rest, he was put on instructing, but was a nervous wreck for flying of any kind.

Chapter 2

The Flaming Rumpler

IN THE FIRST world war many of us who have since flown the international routes of air transport had our first thoughts about the airplane being used as an instrument of peace and human relations rather than one of war and destruction. For me these thoughts culminated suddenly and dramatically in a decision over France on a high-summer day of 1917.

I was flying a little Sopwith Scout, a single-seater fighter of those days with an 80-horsepower LeRhone rotary engine. This was a lovely, light, ethereal little creature which could climb to twenty thousand feet and turn inside almost any other aircraft in the sky. But it had serious limitations as a fighting airplane. It was slow, not very strongly constructed, could not be dived at high speed, and the rate of fire of its single Vickers gun, synchronized to fire between the blades of its propeller, was slow and ineffective compared with that of the twin Spandaus on the nose of the Albatros Scout, its opposite number in the German Air Force. The harsh crackle of the Spandaus on the Albatros sounded like somebody tearing a piece of heavy canvas behind your tail.

Unusual tactics had to be used with our little fighter if we were to have any success at all in the destruction of enemy

aircraft. Only by surprise attack could we engage the enemy
because he had the initiative in every aspect of performance.
The Albatros could outclimb, outpace, outdive, and outgun
us by a wide margin in every case: but we could turn inside
him; and so we had a chance of survival, but little chance of
shooting him down in any engagement except in the first
burst of a surprise attack.

I discovered this with some disillusionment when I first
went out to France, with 66 Squadron in March of 1917. But
our morale had been influenced by stories of the high performance of the machines with which the new squadron was
equipped. So we really had our tails up when we flew over
the lines for the first time. Nothing happened that day except
our first experience of anti-aircraft gunfire, which I never
learned to treat with the indifference its results deserved. I
could never get over the uncomfortable feeling of somebody
on the ground secretly drawing a bead on my own particular
airplane and sending off a shell which could destroy me instantly and with absolutely no warning. I was less concerned
about the much more dangerous Hun fighters, who could be
seen and engaged in some sort of equal combat, even with
the inequalities in aircraft performance.

On the second offensive patrol, over the Douai area, I became separated from the rest of the flight; mainly because I
hadn't yet learned the art of keeping formation, and watching
my tail and the surrounding air. I lost the formation in some
light cloud and found myself alone. It gave me rather a delicious sense of freedom; a freedom to hunt down the Albatros
alone and destroy him over his own country, for I still believed that I had a better machine which could easily outclimb the Hun.

Almost to order I had my first sight of enemy aircraft, a
formation of brilliantly colored machines flying ahead and
about three thousand feet below me in the direction of the
line. One, a red machine with green tail, was straggling behind the main formation. The whole situation could not have
been better. I would attack this straggler out of the sun, pull
out, climb away, and sit safely above the Hun formation till I
could get into position to attack and shoot down another aircraft. It seemed just as simple as that.

I came in, apparently unseen by the red and green Hun, tense with the excitement of my first attack. From inexperience, and unable to hold it any longer, I opened fire too soon on the Albatros, who swept round from my line of fire in a climbing turn. I pulled out, opened to full throttle and climbed to keep the height advantage for another attack. I saw the flashing colors of the Hun sweep under the wing of my machine, very close it seemed; and beyond, I could see the enemy formation turning and seeming almost to hang on their propellers, which I afterwards learned was typical of the Albatros when reaching for height. I was excited, but unconcerned about the outcome, secure in the belief that I could easily outclimb the Huns and retain the initiative for attack when it suited me.

But now my Hun was turning in behind me and somehow he had climbed above my level. I pressed the throttle hard forward but there was no more movement. I checked the fine adjustment; but there was no more power available from a better mixture. A creepy horror feeling ran through my body. The Hun had outclimbed me, and now had reversed the situation and was coming in for the attack. I screwed my head round, twisting my neck to try to keep the Albatros in sight. As long as his nose was not on me he couldn't get me in his guns. I waited till he had almost come into position, then pulled my machine round and up in a climbing turn. He flashed by above me and from the corner of my frantic vision I could see the others—all around me now, and above.

I had no time nor thought then for disillusionment about the performance of my machine. I flung the little Scout about and tried to keep out of position from the nose of each attacking Albatros; all the time working towards the lines in any respite from attack. I shot away all my ammunition in snapshooting at the Huns, without chance myself of any concerted attack.

For twenty minutes my little machine danced among those brilliantly colored "Albatri." Holes appeared in the wings but nothing vital was hit . . . till eventually they turned away, only one Hun hanging on and harrying me back to Arras: and when he had gone I sat there drained of all life it seemed, but my little machine still flying for me.

Eventually I began to come back, turned west for Vert Galand Farm airfield, and came in from the experience which changed my whole approach to fighting in the air. That night I lay awake and thought it out. It was going to be altogether another kind of business. Only by cunning and unfailing alertness could I survive and have any chance to destroy enemy aircraft.

And so it went on in 66 Squadron. New pilots came, and disappeared; but some of the originals, cagy and awake to every chance given to them by the one unbeatable quality of their aircraft, its flashing maneuverability, stayed alive and flying.

Surprise was our only chance of victory; and cornering a Hun over our side of the lines: a rare but exciting opportunity. After several months in France, and the promotion of "A"-flight leader to command of a squadron (Major J. O. Andrews, D.S.O., M.C.), I became flight commander.

Coming back from offensive patrols I would sometimes stay up high, in the hope of finding a Hun reconnaissance machine over our side for photographs: and it was on one of these hunts that the whole pattern of my future flying life was set.

We were coming in from the Roulers, Menin, Courtrai patrol one day at fifteen thousand feet. Away to the north over Belgium and heading west over the lines, I saw two specks in the far distance. We had acquired an instinct for identifying Huns, long before we could see the shape of the aircraft, or of course the black-crossed markings. These were Huns: high-flying Rumplers probably, relying on their height and speed to make an unescorted reconnaissance behind our lines. German tactics were different from ours in this respect. The R.F.C. sent over a flight of relatively slow Sopwith two-seaters at middle altitude, upon the theory that the rear gunners could defend the flight against fighter attack. On a flight to Valenciennes, for instance, the formation was certain to be attacked, and probably harried at least most of the way back. Many of these aircraft were lost because in practice they were often forced to turn and dog-fight, the formation would be split up, and the Albatri could then pick them off individually. I remember one day when all six aircraft of 43 Squadron, which also was on Vert Galand Farm, failed to return. Reading from the other side years later, in the life of Baron Manfred von Richthofen, the great German fighter pilot, we know what happened to the 43 formation that day.

But the German Air Force used this high-flying Rumpler, which was faster than any of our fighters except a specially "hotted up" S. E. 5, and relied upon its ability to avoid attack. The Rumpler adopted similar tactics to the R.A.F. Mosquito reconnaissance aircraft in the second world war. He just opened his throttle (at any altitude up to 22,000 feet in 1917) and flew away from the fighters. So there was a tremendous thrill in stalking and getting into position to intercept the powerful Mercedes-engined Rumpler with the little 80 LeRhone Sopwith Scout.

I held my course, watching the two specks to the north, slightly above our level of fifteen thousand feet. A hollow feeling of disappointment came over me as I saw them turn and start to fly back towards Hunland. Apparently they had seen us. They were very alert and cunning, these Rumplers. They had to be, to avoid the very thing we were seeking to do: to catch them over our side so that they would have to pass us to return to their airdrome.

I turned the flight away to deceive the Rumplers as though we had not seen them, and flew in a southerly direction; but watching, and never losing sight of the specks to the north. In a few minutes we got results. The Huns turned again and headed back for the lines, apparently now convinced we had not seen them. I started to climb now, for the height we needed to give us speed to overtake and attack these aircraft, and continued to fly as though we had not seen them, keeping the specks just in view. They were in the trap now, flying steadily on course for Saint Omer. I turned north, climbing at full throttle, and soon had to continue the turn to keep them in sight as the distance lengthened between us.

We were back over our side of the lines now, with the Huns deep into Allied territory, but I was losing them momentarily as their superior speed widened the gap between us. I had to take a chance on losing them altogether rather than use up height for speed. We were at 17,000 now, about on their level, so I continued the climb to 18,000 and held on

towards Saint Omer; but the Rumplers were gone. Then suddenly I saw a glint of sun on a tiny speck in the west, and
was able to hold it as it moved slowly in a northerly direction.

They must have finished their job and were flying north before turning to come back, on a different track. I turned
again parallel with their course. They would have to turn
soon, to come back over the lines. We had nearly 19,000 feet
now, with some height to use up for speed.

I think the Huns saw us now, and realizing that we had
closed in behind them, decided to turn for their lines and try
to run through us on their speed. The gap between us closed
rapidly. Now they were airplanes, the upper surface of the
wings and tail a dull camouflage color. One was well ahead
of the other. I selected the first for a frontal attack from
above, intending to go through, turn, and come back under
his tail if he was still flying.

I was shaking with the excitement of the chase. Everything
had concentrated on this moment. All the tension of uncertainty as we had stalked these Huns let go as I turned and
came in for the first Rumpler. Tracers came smoking close
by my machine and as I opened fire I could see the rear gunner crouching and firing at me while the Rumpler held a
steady course. The black-crossed airplane rushed in towards
me and swept through below. I brought my machine round
for the stern attack, but the Rumpler was already out of
range. I had misjudged the turn: left it too late, and he was
gone. Other members of the flight came through to the attack
and chased him into the east but the Rumpler was too fast,
and eventually managed to escape.

The other aircraft was coming now, about five hundred
feet below. There wasn't much height in hand to beat the gap
in speed. I could not risk another double attack. As the Hun
passed under me I put down the nose in a steep dive, went on
through his level, and with the speed of the dive pulled up
almost vertically from below. As his blue-white belly came
forward to the sights I followed him through with the Vickers pumping out its rounds . . . till my little fighter stalled in
the thin air and fell away for the earth. I let her go till she

had speed for flight again; then eased her gently out of the dive.

High above, against the clean blue sky, my Hun was still flying, quite straight and level; but a red glow like the end of a cigarette shone out of his fuselage. I watched, fascinated, not yet believing he was on fire. Then the black smoke came, trailing like some funereal streamer from the stricken aircraft, staining the blue of the sky with the signal of death for the Rumpler crew. But the German aircraft flew on, still straight and level as though it would pass without falling to the eternity of space. From a wild, exhilarating wave of triumph at this successful end to a long chase, a dull sense of horror came over me. There was something awful and uncanny about this doomed airplane.

I sat, still and suspended, in the cockpit of my machine. Then a black object detached itself from the blazing Rumpler, and fell away for earth; a grotesque thing with loose and waving ends, and I realized it was the rear gunner who had jumped from the death by fire to which my action had condemned him. He appeared to fall quite slowly, passing my machine as though he were almost floating in space; and then he was gone, invisible against the dark and war-torn earth.

This horror drama continued on its way to the inevitable end of machine and men. The Rumpler, now just a stream of black and putrid smoke, slowly put its nose down, to the final dive of its career. I watched it go, followed it down as pieces came off in flame and smoke. Then it seemed to go out, burst again into flames, and finally hit the ground with a great explosion, leaving a cloud of smoke drifting slowly over the land.

High in the air a suspended silence seemed to empty the sky of all life. For the first time I was horror-stricken by the facts of war in the air. I had seen aircraft go down in flames, break up in the air and flutter down in tattered shreds, go down out of control and hit the ground with a shattering crash, but somehow it had all been impersonal. The Albatros was not, somehow, an airplane with a man in it, but a sleek and dangerous creature of the air, to be destroyed or to be avoided by all the cunning that experience taught us. For some reason this Rumpler was different, or perhaps I had been in France too long. But this now was no triumph. It was a horror from which I wanted to fly away.

The other machines of the flight came in from the sky and slid into formation behind me. It was past the end of the patrol now and we were low in fuel. I turned into the west for Estrée Blanche, our new airfield near Aire, and flew on in the still quiet of the summer sky.

I had known for some time, instinctively, that I was now committed irrevocably to the Air. In this lovely ethereal little airplane I had found a medium of expression which was entirely satisfying. Living with it in the air there were no problems; none of the discord of life on the ground. Here was perfect harmony and peace; but alive, with a complete sense of spiritual freedom. Here, as it is upon the top of a mountain, or on the sea, we were close to God. But why did it have to be used only for war, for killing instead of creating? Why, with this God-given thing, did I have to kill the German crew of the Rumpler?

Returning from this encounter I had no more taste for war. The spirit of the chase had ended this time in desolation. I began to think my way over the world, to my home in Australia; to Lion Island with my boat moored off the beach; the tent by the banksia trees, the red gums sprawling over the sandstone rocks, the call of the penguins coming in from the sea at night . . . Over the oceans, and the continents, from this war in France.

Flying westward in the stillness, I fancied myself going on with the rhythm of the LeRhone spinning a way of life around the world, a way of peace and understanding instead of a way of war and destruction. Three hours' endurance. Three hundred miles' range. Not long ago it was a feat for Blériot to cross twenty miles of the English Channel from Calais to Dover. If my airplane could fly three hundred miles it must be possible some day to fly three thousand miles to join the continents across the oceans. If people could be brought together by fast and easy communications, and perhaps even travel between the continents by air, they would learn to know each other: and if they knew each other personally war would be a personal thing, to be avoided.

The first inspiration was there; rooted to grow into a life's incentive. From school I had been destined to go on to study medicine. Now I knew I would never do that. I could see only an aircraft heading out over the ocean. Living this dream, I had taken the flight too far. It was more than time to descend. I waggled my wings, eased back the throttle, and poured my airplane down the height for Estrée Blanche.

There was some excitement at the squadron. Word had already come in that the Rumpler had been seen shot down, and the wreckage was lying near Brielen. As far as I can remember this was the first Hun we got on our side. Most of our flying was on offensive patrols on the German side and the Rumpler hunts were usually private affairs, mostly undertaken on the way back from patrol or on long personal ventures in our own time. So a Crossley tender was turned on to go up to the crash. I was still sick with the whole thing but, somewhat inconsistently, I went with the others in the tender; mainly to avoid the embarrassment of explanations.

The big Mercedes engine had dug a great hole in the soft earth. The rest was unidentifiable wreckage. A group of tin-hatted soldiers were standing around. One of them came over to me and said, "Want to see the bloke? He's under that sack."

That was the end for me. I turned away.

Back at Estrée Blanche later somebody gave me as a souvenir the Rumpler tail-skid collected from the wreck. I had no wish for it, but to avoid explanation I took it and afterwards gave it to the squadron equipment officer.

Chapter 3

The Way to the Oceans

NEAR the end of the war I was posted, temporarily, on loan to the Australian Flying Corps as fighting instructor to the A.F.C. Central Flying School at Point Cook, near Melbourne. The opportunity to return home for a few months appealed to me, but the war ended soon after I reached Australia and after the first wave of relief there was of course the unexpected problem of peace. War, and life in the R.F.C., had become normal to me, and since we never had any contact with the progress of the war on its higher levels nor, in fact, any real thought of its ending, the impact of peace was a rather surprising phenomenon.

So I quickly seized the first opportunity to make a flight which to some extent was in line with the life I had been living. Somebody bought two service de Havilland 6's for civilian commercial flying and these had to be flown from Point Cook to Sydney, a distance of about 500 miles by the route round the mountains. Though today this is a normal flight for any reasonably experienced private pilot, it was something of an adventure with a D. H. 6 in 1919.

Since I could see no future for my ideas of international

exploratory flight by staying on in the Air Force I arranged for demobilization and got the job of flying one of these airplanes to Sydney. It was at any rate a move towards the use of the airplane for communication and transport, even if only a single flight. It was, however, the first postwar flight from Melbourne to Sydney, and it was important for the people who planned to operate these airplanes for local passenger flights, that they should reach Sydney intact and be received with acclamation.

The D. H. 6 was designed for elementary instruction. It was so slow and so virtually impossible to stall because of its wing characteristics and loading that people going on from it to a normal airplane had to learn to fly all over again. Because of the high, convex wing camber it was known to us as the Clutching Hand.

It so happened that on the day of departure for Sydney there was a very strong northerly wind blowing. The airplane rose off the ground with a run of a few yards into this wind and proceeded to climb practically vertically over the airfield.

The Clutching Hand reached one thousand feet and I began to get mentally adjusted on the course for Benalla, the first proposed refueling stop. But, looking down, I found the airfield still there, and, watching the hangars near the shore of Port Philip Bay, I was somewhat shaken to find them moving the wrong way. The airplane, heading north, was moving south out over the wide waters of Port Philip. At full throttle we were still losing ground; so without further experiment I pushed the nose down to increase speed and just managed to connect up with the leeward edge of the field and put the airplane down. Thus ended the first attempt to make the flight to Sydney.

Ultimately, after two engine failures involving repairs, and the elimination of the second airplane after a series of forced landings due to mechanical failures, I reached Sydney in ten days. But one of the passengers, the representative of the purchasing company, discovered en route, wisely I thought, that urgent business matters made it necessary for him to proceed by train.

The pilot of the second D. H. 6, Lieutenant Oakes, joined me when his airplane was finally grounded, and we made the rest of this epic ten-day flight from Melbourne to Sydney together. However, the arrival at Victoria Park Race Course at Sydney was lit with a blaze of publicity and speeches, and we found ourselves immersed for the first time in the confusing repercussions of such glory.

Through the years following the war I went on flying as a commercial pilot. I also did an engineering course; and I worked in de Havilland's in England for experience. This was an unsatisfactory period, the time when flying as a profession had few stable opportunities and only those of us who found in the air some fundamental and satisfying source of personal expression persisted with it. The freedom we found there seemed also to antagonize some less fortunate people, for I was frequently told by self-appointed advisers that I should "give up all this flying" and go in for some responsible job: presumably shutting myself up in an office, and, from my point of view, ceasing to live.

Beyond the natural attraction of the Air there was always the objective, still over the horizon, of the flights to join the continents. The Australian airman, Harry Hawker, had nearly succeeded in flying the Atlantic, in mid-May, 1919, falling short of Ireland from Newfoundland by only a relatively few miles when the cooling water circulation failed in his engine because of some accumulating obstruction in the radiator. He sighted, and landed alongside, the Danish vessel *Mary*, a small steamer without radio. In heavy seas, he was rescued by a lifeboat from the *Mary* (Captain Duhn) and, after having been given up for lost by all except his wife, was landed a few days later with his navigator, Grieve.

The American flying boat N. C. 4 (Lieutenant Commander Read), one of three which attempted the Atlantic flight, had crossed successfully by the Azores, landing in Lisbon on May 27, 1919, for the first actual crossing. And Alcock and Brown had made the first successful nonstop flight across the Atlantic, from Newfoundland to Ireland, in June 1919. But the design and construction of aircraft, and particularly of engines, had not yet progressed to the really practical stage for transocean flight.

I, quite wrongly as I afterwards discovered, considered myself competent to fly an airplane in any circumstances now; but, like most pilots, I knew virtually nothing about serious navigation. So I sought the opportunity to learn the theory of astronomical navigation. In this I was lucky, because I found a teacher who, rather than impress upon me the difficulty of it all, so enlightened me upon the fundamentals of spherical trigonometry and all that followed it, with such intelligent simplicity, that from being a mathematical failure at school I really became very good at it indeed, and amused myself at home setting myself the most tricky problems to see if I could solve them. Mr. F. G. Brown, my instructor, converted a subject of boredom and confusion to one of clarity and inspiration, and set me off reliably on course to become an air navigator. After I had learned the theory I began to think about instruments.

Very little real navigation had been done in the air. Though a very few of the earliest flights had been navigated with precision and a real knowledge of the subject, in most cases the airplane had been kept on courses worked for the pilot before departure, or just pointed at the broad front of a continent, to be finally guided in by radio; or both. I hadn't the temperament to accept such hazards.

So, having learned something of the theory, I started work on the design of a drift sight which would accurately measure drift of the aircraft over the ocean caused by crosswinds, and on the conversion of a marine sextant to a spirit-level attachment which would enable me to take sights of the sun and the stars, at any altitude of the aircraft, without using the natural sea horizon of the maritime navigator; and thus by laying down and calculating the resultant position lines, would enable me to find the position of the aircraft, and whether or not it had drifted from the desired track to its objective.

With the help of an aircraft engineer and an instrument maker at Sydney, both these projects materialized and I had then the essential tools of my new trade, beyond the magnetic compass upon which all navigation was based; and the chronometer, the accurate timepiece which is a necessary basis for calculations.

The thing now was to learn how to use these tools; how to

convert theory to practice, so that I could find my way over the ocean without the then somewhat temperamental aid of radio direction. For this I had also to learn to identify the stars I needed for navigation. Looking up at night to the thousands of stars in the heavens, this threatened to be a formidable undertaking; but it fascinated me, this seeking out into space to find the way in the air. It suggested a new freedom beyond even that of transocean flight, a first dawn of contact with a vast unknown to be explored as part of that flight.

The textbooks referred me to star charts, mechanical starfinders, to a study of the constellations, of animals and objects said to be outlined in the heavens by the patterns of the stars; and much space was devoted to calculations of their heavenly latitude and longitude. Very soon I was hopelessly confused by all this, and a fascinating subject which had appeared formidable enough in the beginning had now been blown up into one of inexplicable complication with millions of little points of light blinking at me from the night sky.

But at this rather critical stage of my air navigation studies and experiments I happened to meet the captain of a ship which had been lying for some time in Sydney harbor. I exposed my problem to Captain Hugo and he invited me aboard his ship for a session with the stars.

With all the textbooks, star-finders, and other complications put away behind me, we went out on deck and looked into the heavens. It was a brilliantly clear night in December when so many bright stars are visible, and I quickly began to gain confidence from this direct and natural approach to their identification.

Captain Hugo started with the Orion stars. I remember the first one he showed me was the red one, Betelgeuse; then the brilliant white Rigel: and the three little stars of Orion's belt pointing to Sirius, the brightest of them all. Then over the heavens to Canopus, Achernar, and Fomalhaut. Back again to the equilateral triangle formed by Sirius, Procyon, and Betelgeuse; and down to the twins, Pollux and Castor.

And so I began to *see* the heavens, not as a confusion of stars in some way connected with a greater confusion of textbooks and other complications, but as a vast and intimate revelation which I was beginning to understand, and with which I already felt the first touches of an immensely satisfying contact. It was the dawn of a security in the air coming to me from the infinity of space, instead of from a check back to objects or communications on the Earth.

After one other night, shortly before the dawn when more stars had come over into the visible heavens, I could identify enough for practical navigation. The few bright ones remaining in far northern celestial latitudes and not visible from thirty-four south at Sydney would come to me at some future time of flight into the north. I put away forever all artificial aids to identification, and began to feel the joy and the thrill of personal, intimate knowledge of my friends the stars. Though I was not fully aware of it at the time, I had made a very important move towards that essential outlook for an air navigator: a sense of security in space when all earthly contact has gone. One learns to live with stars and space, with confidence in being able to fix the position of the aircraft and, as in a sailing ship wide out in the ocean, one feels secure in being far from any leeward shore.

The first sights of sun and stars taken from the lawn in front of my house with the improvised sextant turned out quite well, and the thrill of laying down star position lines which intersected at a point within reasonable distance of the true position was an exciting and satisfactory conclusion to a great deal of work. But I still had to discover whether it worked in the air; whether I myself could make it work, and whether this sextant was fundamentally capable of producing accurate results in give-and-take conditions in an airplane.

To carry the experiments on to the next phase I bought a de Havilland Moth twin-float seaplane. I had never flown a seaplane but had heard that the take-off was only for special beings called seaplane pilots; that only by some sort of involved technique denied to ordinary mortals could a waterborne airplane be persuaded to become successfully airborne.

Firmly believing this to be true, I made various inquiries of people who at some time or other had flown seaplanes, and with natural caution checked the advice of each expert against the others. All differed in some important respect. The only point upon which agreement was unanimous was that without this special knowledge of how to take off from the water, disaster was certain.

I was so confused by all this that I took the seaplane out on a good open stretch of water and, ignoring all conflicting advice on the subject, steadily pressed the throttle forward and went with the airplane, which I believed knew more about it than I did. It ran a reasonable distance and with light backward pressure on the control column became airborne quite smoothly and definitely. There were, I later discovered, some refinements to this procedure, to meet different sea conditions and the demands of different airplanes to them; but on the whole the thing turned out to be governed by the normal laws of behavior of a flying machine.

The operations of this seaplane had to be made to pay, to cover the economic facts of life: so I flew it to remote lakes and harbors where there were enough people to provide some traffic for passenger flights. I did the maintenance myself and, to economize on personal expenses and to live as I liked, I camped with my dog on the edge of accessible but, as far as possible, uninhabited lakes. A large Alsatian, he was the most good-tempered fellow, with traditionally good manners, but his formidable appearance was a strong deterrent to anybody with ideas of interfering with the airplane if I happened to be away fishing.

He loved the flying and, like all dogs in a car, liked to put his head out the side into the wind from the front seat, in which he traveled; but finally he would get disgusted with the force of the airstream and would just curl down in the seat and go to sleep. The only trouble I had was to teach him to jump out of the cockpit onto the plywood walkway alongside the fuselage and not to put his feet through the fabric of the wing when we came in to anchor. He always wanted immediately to jump out, and either swim ashore or go off on some enticing personal expedition the instant the floats touched the sand on a secluded beach. But he learned to put his feet in the right place on the wing and never did any damage.

One of the most beautiful places where I camped with the seaplane was at Swan Lake, then a quite uninhabited region on the south coast of New South Wales. I had my tent there and would return in the evening after doing passenger flights at Sussex Inlet, Moruya, or other river settlements on the coast.

One evening when I returned to the Lake I was surprised to see two men with shotguns making their way along the shore to a large flock of swans which had come in to camp for the night in a sheltered corner. Their purpose was obvious: so I quickly started up the engine, taxied out of the creek by my camp and, with the engine just ticking over, gently headed the group of swans away from the shore, using the little seaplane like a sheepdog, till I had the swans far across the lake and out of range of any danger.

I soon had enough money to enable me to interrupt my passenger flights and, still living a good deal away from the stresses of civilization, I was able now to launch out over the ocean on experimental navigation flights. The drift sight produced good results. I found I could measure drift quite accurately even flying alone; but the first attempt at sun sights with the sextant horrified me. When I came in and worked them ashore they gave the most erratic position lines, far outside the standard of accuracy I needed. I found the sextant bubble or spirit level too sensitive to movement of any kind, and my own skill in handling the instrument in the air much in need of practice and experience. So I worked out a different kind of bubble chamber designed to a shape which I believed would allow the liquid to balance out accelerations of the instrument, and had it adapted to the sextant. Then I went back into the air to try for some better results. Though it was still obvious that I needed more practice, the results with the new bubble attachment were encouraging, and, allowing for my inexperience in handling, I began to see that we were on the way to practical astronomical navigation in the air.

For the night flights there was always the problem of a flare path, however primitive, for landings: so I decided to try to evolve a system of night landings on the water without flares, which was virtually an instrument-landing system.

This was very simple, but reasonably effective. It was based on a spring-loaded rod which in normal flight was mounted fore and aft in the line of flight under the horizontal cross-bracing struts between the floats. When approaching to land

I would release the rod from its rear strut attachment and, pivoting on its bearing on the front strut, it would be brought down into a vertical position by the spring attached to the forward end, and would extend some six feet below the floats.

By coming in low on final approach with a very low rate of descent and constant airspeed, the airplane settled gradually towards the invisible surface of the water. When the keels of the floats were about six feet from the surface the vertical rod began to cut into the water, the pressure of which began to pull the rod back against the spring. This backward movement, which could be caused only by contact with the water, was registered on a lighted instrument on my panel by suitable connections and it was immediately obvious that the airplane was close to the water. As she continued to sink towards contact, more of the rod would be in the water, an increased reading on the instrument would be shown, and the airplane could be eased into a level attitude, till it just sat on the water. This was a real "Heath Robinson" contrivance; but it worked, and I made a number of blind night landings with it on the training and check flights for myself and the sextant.

After operating this little seaplane for more than a year, doing charter and passenger flights with it on the east coast of Australia and across Bass Strait to Tasmania, and practicing navigation, I had proved that my instruments worked and I believed that I had reached the stage of experience where I could confidently navigate an airplane across an ocean.

It was just about this time—specifically, on May 31, 1928—that Kingsford Smith, Ulm, Lyon, and Warner left Oakland, California, on their historic east-west flight across the Pacific in the Fokker FVIIB–3M aircraft, *Southern Cross*. This great flight made me impatient to proceed with my own plans. But financial considerations delayed them and this proved to be the best thing which could have happened to me, because I was to learn a type of flying which would put me years ahead of the times and right into the technique of airline flying as it is today. It showed me again that I still had much to learn as a pilot, as well as a navigator.

Chapter 4

Airline Wings

AFTER their Pacific flight Kingsford Smith and Ulm had formed the original Australian National Airways and, with Fokker replicas of the *Southern Cross* built in England by A. V. Roe and known as the Avro 10, had inaugurated services between Sydney and Melbourne, Sydney–Brisbane, and afterwards Melbourne–Hobart. Right from the start the spirit of this airline was the spirit of the Pacific flight.

After the new, wide world of the Pacific crossing Kingsford Smith could see no problems on the Sydney–Melbourne run. I later understood his view about this; how, from the far, unearthly spaces of transocean flight he had acquired a sense of immortality, an ease of spirit and a smiling confidence which carried him on to encompass without question the lesser spaces of the mountainous, instrument flight on the Sydney–Melbourne run without radio aids or communication, weather forecasts or terminal reports. It was a trap, of course: but his great, free spirit could see no problems in such flight.

The days of contact flying, following railway lines and roads, of checking from waterhole to waterhole, of cancelling

for weather, were gone forever. The A.N.A. Fokkers went
out on time from Sydney airport and flew as the airlines fly
today, climbing to altitude on course, leveling off and taking
the weather as it came—but with no radio and virtually no
up-to-date knowledge of the weather either on the route or at
the terminal.

Extraordinary people were needed to undertake such
flying, because in spite of Smithy's transpacific smile, the
Sydney–Melbourne run, and in some circumstances the
Sydney–Brisbane, are still quite tough even with all the modern aids of flight instruments and radio, up-to-the-moment
forecasts and reports, and pressurized "over-weather" flight.
With the old Fokker doing exactly the same type of route
flying, clawing its way for height in cloud, often loaded with
ice over the Bogong Peaks, its few gyro instruments driven
by a venturi out in the icy airstream, with no possible way of
fixing its position except in the mental arithmetic of its captain, a rather special approach was necessary for the job of
flying for A.N.A.

The idea of flying for this airline, the first in Australia to
introduce modern airline flying, appealed to me; so when a
vacancy for a pilot came up I applied for the job.

At an interview with Charles Ulm, who was executive
managing director of the company (Kingsford Smith was not
sufficiently earthbound to be imprisoned for long in an office) I was left in no doubt of Ulm's opinion of my qualifications as a pilot for A.N.A. He dismissed at once any question
of my engagement as captain, but offered me a trial as second pilot more or less on probation. This ruthless dismissal of
all the experience I had had up to that time rather shook me
and I was on the point of turning down his offer with suitable
comment when something stopped me. There was something
about this man I liked. He was ruthless and tough but there
was something good about him. He would be equally tough
with himself, and there was a swaggering but genuine gallan-
try about him: and his offer was a challenge. I decided to accept it.

I was put on the Melbourne run as second pilot. Jimmy
Mollison, who had joined the airline six months earlier, was
the captain. I turned up at the airport and stood by the air-

craft to introduce myself to Mollison, whom I had never met. He arrived a few minutes before the scheduled departure at eight o'clock in the morning, passed some cynical observation to me and went on up to the pilot's cabin and into the port seat. I climbed into the starboard without comment and waited to be told what to do. Mollison completely ignored my presence, started the engines and, as soon as they were warm enough, began to taxi out. I didn't think I was going to like this very much but made up my mind to accept whatever was coming and just try to do my job.

There was nothing to do. With a thunderous snarling roar from the three Lynx engines Mollison took the aircraft into the take-off.

I was very impressed. Compared with anything I had flown, this was a gigantic airplane. The whole approach to it seemed utterly different. I had always flown light machines; taken them to me in my hands and flown them. This great monster was flying me. I was just going with it, virtually a spectator apart even from the fact that I had nothing to do. Without knowing it, I had made the first step towards understanding a big airplane. You don't fly them. You go with them, giving them a guiding hand, reacting to their demands and wishes and acting in their interests. They know far more than any pilot will ever know, yet you are indispensable to them. You know they need you, vitally, for the exact and accurate reaction to their needs: and so you become sensitively part of the great patient monster you control, usually lightly, sometimes firmly, very occasionally decisively, but never with brute force against the structure of their bones. With perfect balance of air forces and weight they can be feathered onto the land or water like a single-seater scout, but it has to be right from the beginning: a smooth, progressive operation beginning with entrance to the circuit and ending with the transfer from air to earth.

But I am moving too far ahead. I sat on in the starboard seat while Mollison climbed out on the course for Melbourne. At 8000 feet (no odds and evens then) he leveled her off in clear air, eased down the power, synchronized the fixed propellers with the throttles, and sat back with a disdainful air of boredom. I thought back to the interview in Charles Ulm's office and was glad I had accepted the job. There was something new in all this, different from any flying I had done.

We were nearing Canberra. Ahead, a great buildup covered the mountains, the overflow from solid cloud building in on a southwest wind from the other side. The tops were thousands of feet above our level, and the base down in the treetops on the lower mountains even before the Bogong Peaks. I didn't like the look of this. I glanced out to starboard and ahead to lower country round by Yass and Gundagai. Soon Mollison would realize there was no way through and would descend and go round by the western plains. But he showed no signs of altering course; just sat there with his hands quite gently on the controls, with the great swirling mass of turbulent cloud closing in towards us, reaching to the heavens, and down into the rugged country beyond Canberra.

I realized that he was not going to descend: that he was deliberately flying the airplane into this mass of blind cloud probably extending over the mountains to Melbourne and with nothing to tell us about the weather conditions at Essendon airport. To me it was a grim and hazardous prospect which I frankly found difficult to believe was being voluntarily undertaken by Mollison, with his roaring Fokker already flying in freezing, menacing conditions.

To understand and to appreciate this situation it is necessary to know that in Australia this had never been done before; that the science of instrument flight and dead-reckoning navigation without sight of checkpoints on the earth was relatively new in aviation, and where in the United States and Europe it was regularly undertaken it was based upon radio communication and aids to navigation, organized weather reports and forecasts, and already some form of anti-icing and de-icing protection for the aircraft. Without these aids and safety provisions, regular "in cloud" and "on top" flying could be undertaken only by freak pilots with a new and wider perception of flight, a relentless cunning, and a rugged ability to handle the airplane on primitive flight instruments in the most violent turbulence. They had to be sensitive, intelligent, with imagination which could project itself forward to situations well ahead of the aircraft, and therefore people capable of fear: yet they needed also the steely quality, the basic strength of character which would retain stability of action in any circumstances. This combination was extreme in Kingsford Smith and was largely responsible for his greatness as an airman. The conventional conception of the strong and silent he-man, fearless and therefore upon whom impressions do not register, is a very great menace in the air. His days are numbered from the start. So also is the supersensitive but uncontrollable imagination which allows dismay to rule and panic to take charge. Kingsford Smith had imagination but he was not dismayed by his imagination. The pilots of A.N.A., each a completely different individual, each a phenomenon in his own way and blatantly a genius at his art, all had this combination.

Mollison, beside whom I sat with considerable misgivings heading into this impossible weather, was almost effeminate in his manner. I think it was partly a slightly mischievous pose, partly natural: but he was a dangerous man. There is an authentic story of a night at a dance in Melbourne. Some semi-drunk was foolish enough to mistake the meaning of Mollison's waved hair and to insult him in front of the lady with whom he was dancing. Mollison quietly took the man outside, asked him his address, got it: then beat him up into insensibility, called and paid a taxi, and had the remains delivered to the address.

Some years later, when transatlantic adventure flights in small airplanes were preceding the big four-engined transports of the regular services, Mollison decided to fly the Atlantic, from New York to London. For this venture he had, tanked up to an outrageous overload weight and fueled at the airport, a very temperamental airplane. This was a single-engined landplane which only a Mollison would have taken on. After some delays for weather and other influences Jimmy Mollison was dancing one night with a lady of whom he had seen a good deal in New York. They went outside. It was a beautiful night with all the stars in the sky. Mollison looked up, thought for a moment, and remarked casually, "I think I'll go, my dear."

"Go, Jim? Where?"

"Oh . . . to England, you know."

Whereupon they hailed a taxi, drove to the airport, and Mollison, white tie and tails, got into his airplane and flew it 3000 miles, alone across the Atlantic, to London, and landed at Croydon airport.

But I knew none of this of course on the Melbourne run that day. I wasn't at all convinced that Mollison would even be able to control the airplane in the turbulent cloud ahead; and the chance of descending without hitting a mountain, and of finding Essendon airport, on which my imagination was already working overtime, seemed virtually nil to me. I thought of trying to persuade him to descend and go round, but I instantly discarded this idea because I knew that A.N.A. had risen out of this type of flying, hopping from twig to twig; but I still couldn't believe this was normal, or that it meant anything but disaster.

And then we were in the cloud. The Fokker hit the swirling air in the ragged, wind-blown edges; and reacted with a bouncing, snarling roar; and sight was gone. I saw Mollison's knuckles tighten a little on the wheel, but beyond that he showed no visible reaction. The needle of the turn and bank indicator swayed and jerked from side to side as the turbulent ocean of air sought to turn the Fokker away and force her out of control. The airspeed rose, and fell, as these forces tried to send her nose up towards the stall, then down for the cloud-shrouded mountains.

From the corner of my eye I watched Mollison. Was this just mad bravado, or did he really know how to do it? I watched him very closely, for this was when I would know the answer. All his actions were deliberate yet somehow coordinated as he clearly interpreted the movements of the aircraft through the faces of the instruments. In a few moments he had her settled down, bouncing and snarling still, but under control, pressing or drawing the control column against the combination of his senses and the movement of the airspeed indicator; holding on aileron and rudder against the movement of the turn and bank. The whole process settled down to a rhythmic, steady advance through the boiling air, with man and machine in perfect unison and understanding. The Fokker told him, through the instruments and his own reactions, what she needed from him, and he responded and supplied those needs. There was no doubt about this. Mollison had control of the airplane. I relaxed a little into my seat and began to speculate upon the future.

Between us and Melbourne airport was continuous hilly and mountainous country, till the last twenty miles of flat and undulating land. Beyond were more scattered hills and Port Philip Bay. Somehow Mollison would have to get this airplane down out of this situation without flying into a mountain, and he would have to visually locate Essendon airport.

This was a disturbing prospect. How would he know the wind? What was the drift; and our groundspeed? If no sight of the land appeared and we could not identify our position, how would he know when it would be safe to descend? I didn't see how he *could* know: and I wanted to know, infallibly, before those throttles were touched and any attempt made to descend. I was becoming hostile now as I saw and weighed up the possibilities after the initial doubt about control had been dispelled by Mollison's obviously competent handling of the airplane.

But did he really know where he was going, or was he just chancing it, hoping for the best, and pulling off the throttles when he imagined he ought to be over low land? Suppose he even allowed a wide margin, beyond Essendon? He might collect one of the hills out towards Geelong, or, with unknown drift on the westerly, end up in the mountains beyond Dandenong. Whichever way I looked at it I didn't like it.

I looked out to the starboard engine and to the great Fokker wing driving on into the gray mists of cloud. Ice was forming on the struts of the engine mount, and on the leading edge of the wing. I began to remember and think of the passengers in the cabin. They could know nothing of this. They had presented their tickets at the airport, taken their seats, and were going to Melbourne. A merciful gulf separated them from reality, though not from the results of it. I again let my glance observe Mollison. He was flying with one hand now, completely expressionless.

We plowed on through this weather for three hours. I kept constant watch out to starboard for sight of the land. Twice a gap, a deep hole, appeared, and once a shadowy region of stratified cloud with trees moving under a veil of mist. But we were in it without hope of making a visual descent.

Five hours and forty minutes after leaving Sydney, Mollison announced casually, "I think we'll go down now." He drew off the throttles and let the nose go down. My thoughts immediately projected themselves ahead; but what was the use? We had to go down somewhere. There was no sight of the ground and no prospect of it, apparently. Added to the completely blind conditions, heavy rain was now screeching against the windshield and streaming back past the cabin window, oozing in through cracks in the cabin; dripping around my legs. The whole thing was completely fantastic; beyond further worry. I resigned myself to the inevitable as Mollison took the Fokker, grumbling and snarling, down through the ocean of cloud.

I watched the needle of the altimeter chiefly, and outside for a gap and sight of the earth. Even one gap would show me whether we were over low land or still over the mountains; but there was none. We went on descending, into increasing darkness and a mixture of streaming rain and spurts of blacker cloud.

Four thousand on the altimeter now and still no sight. I had gone over all the heights and distances on the Melbourne run the day before and knew that the ranges north of Melbourne were up to three thousand feet. Soon we would have to be past these ranges and over the low land or we would collect a hillside in this absurd descent.

He kept on just enough power to keep the engines warm and continued on down on the course with unvarying rate of descent. At 2000 feet I had more or less made up my mind that a crash was inevitable, but I hoped that when the trees appeared out of the cloud in front there might just be time to open the throttles and climb away; but actually I knew there wouldn't be, unless a miracle happened.

At fifteen hundred feet a few broken wisps went by with

gaps and I saw flat land, and survival, below. At 800 feet we broke through into the clear and I saw Yan Yean reservoir moving away astern. We had come out fifteen miles from Essendon. In ten minutes we were on the airport.

That evening in Menzies' Hotel I again thought back to the interview with Charles Ulm. How right he was. I was completely incompetent to be captain of such an airplane on the A.N.A. runs. This was a new kind of flying. Could I go on, and learn to do it; or had I, as it had been suggested to me by friends, reached the age where I was too old for it? I was thirty-four; getting on I suppose (according to the general ideas of those days); but why? There was nothing else I could not do at thirty-four that I could do at twenty-four. Why not fly an aircraft? Boumphry, an ex-cavalry officer, had been thirty-two in my flight in France. We used to joke with him about his venerable old age for flying. And Mollison. He was younger than I certainly, but if he could do it, and the others, why couldn't I? Mollison certainly had an aura round him in that airplane, but he was human; though I couldn't see how he had known when to descend for Essendon.

We flew back to Sydney the next day. The godlike Mollison relaxed and told me. About ten minutes before he shut off to descend he had had the luck to identify a sight of the ground through a passing gap below the port side. He knew his position and it was easy then to estimate his arrival at Essendon and therefore when to start his descent. Suppose he had not seen the gap and identified his position? Well; he did see it, so ... the A.N.A. runs went on.

I flew for three months as second pilot with A.N.A. Then Charles Ulm came to Melbourne with us one day. I had to fly the aircraft through a lot of turbulent cloud. In Melbourne I rather stuck my neck out and asked Ulm how he thought I was going. He said, "I've never been more frightened in my life." But the next day he promoted me to Captain and I flew as an airline captain with A.N.A. till the unsubsidized company ceased operations after the financial losses incurred by the tragic disappearance in March 1931 of the aircraft *Southern Cloud* on the Melbourne run, and the great

world depression, from which Australia did not escape.*
Among the wings I have worn since those of the R.F.C. in
1916, I value very highly the badge of the blue overall suit
with the words, "A.N.A. Pilot-in-charge."

---

* Twenty-seven years after the *Southern Cloud* set out for Melbourne and flew into the violence of a storm, her remains were found
high on the side of an inaccessible mountain about two hundred miles
northeast of Melbourne. A man exploring this rugged country on a
survey for the Snowy Mountains water conservation scheme walked
right into the twisted steel tubing and engines of an aircraft which was
positively identified as the *Southern Cloud.*

Chapter 5
Trans-Tasman, 1933

THE PERIOD as an airline captain with the original Australian National Airways was a most satisfying experience. From the obvious limitations in experience which I had as a pilot when I joined A.N.A. as a first officer and which were most drastically illustrated to me on the first flight to Melbourne with Mollison, the demands of primitive instrument flying in all conditions and those of making reliable airline flights in all weathers without the aid of radio direction or communication, or any organized aviation weather reports or forecasts, had in quite a short time given me all the pilot handling experience necessary for the future.

So I sought out an opportunity to apply my experimental navigation to actual transocean flight where, to bring the aircraft in to her destination, the results just had to be right.

There had been some opportunities, off the regular run with A.N.A., to gain further practical navigation experience, but none of these involved astronomical methods. I had taken out one of the 3-engined Fokkers for the company on a long charter through western Queensland, where we had to find isolated homesteads in featureless country and land in open paddocks; and I had taken a little Percival Gull across the Timor Sea to Batavia in the Netherlands Indies to bring back hurriedly the latest pictures of the England-Australia cricket match for the newspapers and the newsreels.

But none of this had the uncompromising commitment of long-distance flight over the ocean: so when Kingsford Smith announced his intention to make a flight across the Tasman Sea from Australia to New Zealand in the *Southern Cross* I saw my opportunity to make a real-life proving flight of my instruments and particularly of myself as a navigator.

A navigator was needed for the *Southern Cross,* so I saw Kingsford Smith and we had a talk on the various aspects of the flight, including my own inexperience in transocean navigation. It was agreed that he would let me know the following day. At the appointed time he had made up his mind and the answer was brief and "yes." And so with the utmost simplicity we entered into a flying partnership which was to affect both our lives so very much over the next two years. It was typical of Kingsford Smith that from the moment that decision was made he seemed to dismiss the navigation from his mind and he never at any time questioned my work, though there must have been times when he might well have wondered how it was going to turn out.

For the Tasman flight we had my drift sight and improvised bubble sextant, a very good Waltham chronometer, the necessary astronomical information from the Almanac and, as a basis for the position lines of astronavigation, a little book called *Johnston's Cloudy Weather.* This was really the first sight into today's methods of working and using position lines. It was devised and used by a certain Captain Johnston to determine his intersection of a line up which he could turn to put his ship on course for the English Channel when he was homeward bound from the South Atlantic. With the aid of *Johnston's Cloudy Weather* and the rest of the equipment, I managed to hit off a point on New Zealand which was south of our objective by exactly the amount by which I had refused to believe my own navigation, and the magic results of twelve hours' close concentration upon the navigation of an airplane were for the first time revealed to me in stark and surprising reality.

We left Australia at 0300 on January 15, 1933, from Gerringong Beach, a stretch of curving sand which provided a long enough run for the overloaded *Southern Cross* to become airborne. It was a black dark, humid, and eerie night with a misty northeast wind blowing in from the sea. As Kingsford Smith headed the aircraft into the take-off a few flickering lights ran by close under the port wing and glistening foam from the surf reached in for the starboard wheel. But far down the beach the great Fokker wing lifted her away and we slowly turned and headed into the night.

In orderly preparation, I had put the course on the compass for New Plymouth, our objective in the southwest of the North Island of New Zealand, after guessing the probable drift before we left the beach. Now I went below and took a back bearing on the prearranged light. I was immediately alerted with surprise. The wind had been light, almost a calm, as we took off from the beach. Now the back bearing showed twelve degrees of starboard drift. Something was happening which didn't seem right.

Twelve degrees was a lot of drift.

Was my bearing right?

Where would we end up if the whole thing was wrong and we had no drift?

Would we miss New Zealand altogether?

I went back and took another bearing on the fast-disappearing light. Still twelve degrees to starboard. I would just have to put this on the compass, for better or worse. I gave Kingsford Smith the corrected course and looked out into the coal-black night. Not a sign of a star—nothing. Only the roar of the motors and the faint glow of the luminous instruments. Projected thus from the sheltered life of earth, it seemed utterly fantastic that I should be navigating this airplane for New Zealand: fantastic even that there could be any New Zealand; any earth at all in this void of sound and darkness. There was something irrevocable about the bars of the verge ring lined up with the needle of the compass. In this small and luminous bowl was uncompromising evidence of my commitment. If I was wrong, we could end up in the ocean. If I was right there would be the thrill of Mount Egmont ahead, and then New Plymouth.

As we flew on the course I began to absorb the sound, and the vibrant life of the *Southern Cross*. The scattered remnants of the Earthlife, mixing in disorder with the sudden effects of blind but dominant flight as we passed into the night from Gerringgong, were swept away in the airstream, leaving us alone but in harmony with the world of the aircraft. In that world there grew a sense of security: something which did not admit the possibility of an engine failure, or the certain disaster which would follow it with the heavily laden airplane. I acquired the feeling that the *Southern Cross* was set in her orbit, inevitably flying in this dark region of space.

I could do no more about the navigation because, in the overcast, there was no sight of stars or sea; so I let her go on the course reckoned up from the back bearing on Gerringong, and after a while relieved Smithy for a short time at the controls.

As the hands of my watch began to show the prospect of day there was a sense of anticipation again. What would that day bring? What would sight of the sea tell us? There might well have been a complete change of wind. The air had been turbulent, its normal flow disturbed by something so far invisible to us. We might even now be flying in a reversal of wind, being set far off the track to the north with twelve degrees already on in that direction. I thought of altering course, taking off the allowance for drift. At least that way the error would be only that of the drift itself. But, somewhat uneasily, I discarded this idea. There was no evidence to prove a change of wind, only the turbulence and that was not enough. So I said nothing to Smithy, and left the course on the compass. That is one of the satisfactory things about the air. There is usually only a clear-cut decision to be made. Black or white, with no shading in between. If you have any imagination it may continue to whisper in your ear even after the decision is made, but it has to *be* made; there and then on the best facts before you; and acted upon. There is no lying awake at night waiting for conferences; no endless debate and argument; none of these mental hazards which waste so much time and energy in the day-to-day affairs of Earth.

Much of the tranquillity of the air is thus due to the simplicity of the situation in which one finds oneself.

But this morning over the Tasman Sea I wanted more information; something clear-cut and definite to add to the facts, and to check the assumptions upon which we were heading into the darkness.

As the first light of dawn began to dissolve this darkness I watched and waited for a sight of the sea. The first indication of anything beyond the cabin of the *Southern Cross* was a faint lightening in the air around us, showing the aircraft to be passing in and out of cloud. In the gaps we were soon able to recognize the surface but it was strange and intangible with a shiny whiteness that gave no indication of the wind. Then as the increasing light gradually brought up the definition of the region around us the significance of this surface suddenly came to me. The sea was lashed into streamers of white foam by a wind of gale force blowing from north across the track of the aircraft. I moved immediately to the drift sight and was amazed and horrified to make a reading of thirty degrees to starboard. How could I believe this sinister drift? Out there irrevocably over the ocean the bleak and menacing surface stared back at me, coldly proclaiming the seriousness of my lone responsibility. I tried several drift sights and each time the *Southern Cross,* crabbing grotesquely in this relentless stream of air, gave me the same reading on the drift sight. It seemed that she was being blown on the wind, carried like a canoe trying to cross the rapids, but heading to the oblivion of the wastes of ocean beyond the north of New Zealand.

I left the drift sight for the chart table to get some stark figures into this situation. On the way I noticed Kingsford Smith sitting, apparently unconcerned, at the controls; and Stannage, completely detached, at the radio. I resisted a temptation to confer with Smithy, realizing that this was my affair. This was not theory and academic work in the comfort and security of Sydney; nor intellectual discussion about the latest navigation tables. This was stark reality where I either put the correct course on the compass or we ended up in the drink. If I was ever going to be a navigator, this was it.

There was order and precision at the navigation table. I

could reduce the effects of the violence around us to some logical and orderly conclusion, but the result I saw in the figures was startling. Our track from Gerringong to New Plymouth was 104 degrees, True. The Variation was 11 degrees East; the deviation 3 degrees East. To add now the effect of thirty degrees of drift would mean that the new course on the compass would pass more than one hundred miles north of the most northerly point of New Zealand. With the confidence of experience behind me I would not have had a care about this, but to blatantly put a course on the compass apparently heading us out into endless ocean left me uncomfortable and mentally reaching forward to the time of day when the sun would be coming round abeam and I could get a position line to confirm the track of the aircraft; it we were not still in or under cloud.

I will not go through all the subsequent detail of the navigation up to a time approaching noon, where the sky had cleared to broken cloud over which we flew in calm air with evidence of a much reduced wind on the water. It was now that I would know the results of my long discussions with the instrument maker in Sydney, experiments with various kinds of spirit levels, air trials and other preliminaries which had led to the final form of the bubble sextant I now had with me in the *Southern Cross* and which was destined to provide the vital information I needed from the sun. I drew out the sextant from the polished mahogany case in which it had previously lived for many years its life at sea, and opened the protective lid over the rubber-mounted chronometer where the third hand on its face was methodially ticking away the seconds of Greenwich time.

I lifted the sextant to my eye, brought the sun down to the bubble, and held my breath to steady the sensitive reading of the instrument. In a few minutes of intense concentration I had the series of altitudes of the sun and the chronometer times of their observation. Back at the navigation table I averaged these, worked the results, and assembled the information to lay down the position line on the chart. The *Southern Cross* took no heed of this. She rode on the air, smoothly and patiently heading to whatever course I might put on the compass.

The location of the position line was another shock. It was exactly opposite to the direction in which my imagination was putting the aircraft. It clearly showed us to be seventy miles *south* of the track to New Plymouth. Had I made a mistake? Was the sextant accurate? Could it have been damaged? Was there something I hadn't taken into account at all? Kingsford Smith still sat there unconcernedly flying the aircraft. Stannage was with his radio, going to New Zealand. The *Southern Cross* had no knowledge of my anguish. I was alone with this problem. I was the navigator.

I thought, of course, of the obvious reason for this set to the south. It had happened in the night when it was impossible to observe the terrific drift. But I still couldn't believe that on a compass course 44 degrees north of the true track we could be 70 miles *south* of the estimated position. I still felt that there might be something I hadn't taken into account; that perhaps I had made some omission or mistake; that if we were north of the track and not south as my position line showed us to be, if I altered course still more to the north in obedience to the evidence before me and it happened to be wrong, we would be hopelessly lost over the ocean.

To stabilize my own reactions, which now were the real threat to safety, I returned to a cold study of the facts upon which I could infallibly rely. I wanted to hit off New Plymouth on the nose. But to be sure of hitting New Zealand was more important. The figures showed me that if I assumed a position thirty-five miles south of the track instead of the seventy shown by the position line from the sun, we *must* hit New Zealand; unless the whole thing were wrong!

So I set a new course for New Plymouth from a position 35 miles south of the direct track, and estimated our time of arrival.

As we approached New Zealand cloud built up again off the port bow and we could not see Mount Egmont: but ahead and to starboard it was clear, and as the E.T.A.* came up we saw far away off the starboard bow the strain of low land against the ocean. This was identified from the chart as Cape Farewell, the northwest point of the South Island of

---

* Estimated Time of Arrival.

New Zealand. South of the track! Just as the sun line had told me. We were running in to Cook Straight, south of the track into New Plymouth by almost exactly the amount by which I had refused to believe my own sights.

This was a revelation: absolute proof that it worked: that the compass courses in the twelve-hour-long sequence of reckoning and allowances, observations, calculations, and steering of the aircraft would have brought us right in to New Plymouth if I had believed and acted upon the evidence of my own sights with the sextant.

The work of years had been focused into this landfall: all the spherical trigonometry, the long and arduous study of marine navigation methods, the seeking after the little that was really genuinely known of its application to the air, the development of the instruments, and the experiments in the air: everything I knew at this time went into the navigation of the *Southern Cross;* and it was right. That was the terrific satisfaction. Every action affecting the navigation in that fairly long passage across the Tasman Sea must have been sound in principle and right in application but lack of experience had allowed my imagination to convince me that there might have been factors beyond my use or knowledge which could have fatally affected the flight. I had little upon which to congratulate myself, for missing New Plymouth, but the future prospect of navigation across the oceans was intensely exciting and was calling me strongly to the next flight.

Chapter 6

Pacific Flight, 1934

THE PILOT-NAVIGATOR partnership between Kingsford Smith and me suited us both very well at this time. He apparently was satisfied to accept and to steer the courses I put on the compass, without questioning my reasoning and calculations; and because of the very great respect I had for him as a pilot I accepted with complete confidence his handling of the airplane. He was, in a way, the guinea pig for the progress of my experience as a navigator, and we used to joke about this; but as the vital objective continued to come up in the right place each time we both acquired a confidence in the fact that it worked.

We made three more Tasman crossings in the *Southern Cross*, with John Stannage as radio operator. These were experimental flights, some with mail, and were designed to gain experience, and to stimulate public opinion not only to accept but to demand a regular air service across the 1200 nautical miles of ocean between Australia and New Zealand.

After the fourth crossing, in 1934, we made the first west-east crossing of the Pacific Ocean, flying this time in a single-engined landplane, the Lockheed Altair, from Brisbane,

Australia, to Oakland, California, by the Fiji and Hawaiian Islands. At first sight this appeared to be a most hazardous venture, but there were certain aspects of it which, in terms of the long transocean exploratory flights of the times, had acceptable risks and were based upon sound reasoning. Most twin-engined airplanes then would not fly on one engine with the outrageous overload of fuel we had to carry. Except in the latter stages of the flight after fuel consumption had reduced this weight, failure of either engine would mean a forced landing in the ocean. With only one engine there was only half the risk of forced landing from engine failure.

There were no radio aids to navigation because none existed before Honolulu, and in any case we had no receiver to use these aids. We had by this time, however, acquired a good deal of faith in the results of astronomical navigation in the air, and believed that we could find the islands which were our critical objectives, within the time limits of our fuel.

This flight was undertaken as an alternative to the England-Australia air race, won by Scott and Black in a de Havilland Comet, from which circumstances beyond our control had compelled us to withdraw the Altair's entry.

We also had discussed this west-east Pacific flight a good deal and wanted to undertake it to create interest and confidence in the inauguration of a regular transpacific air service, joining Australia with North America, for which suitable aircraft were already on the drawing board. We believed that a Pacific crossing by a small single-engined landplane would inspire confidence in a regular passenger and mail service operated by the large four-engined flying boats which were then projected.

The predeparture period before the Altair flight was normal in its panic and confusion. Over the orderly activities of preparation in installation of extra fuel tanks, arrangement of navigation space and equipment in the small cockpit, and all the rest of it, there was active news drama in our announcement that, having withdrawn from the air race under the fire of criticism by self-appointed experts, we now proposed to fly the Pacific. We carried on with our preparations and tests and were ready to fly to Brisbane on the day before our estimated departure for Fiji.

On the run to Brisbane we made one of those typical horrifying discoveries which are more or less normal with exploratory flights. We found, from the results of our fuel consumption on this run, that the Altair could not fly from Fiji to Honolulu. We had run the engine at power settings calculated for us by experts, to give maximum range; and with that predicted maximum range there was adequate fuel for the longest flight stage. The Sydney–Brisbane run was intended to be merely the formal confirmation of the power setting—fuel consumption figures worked out for us. Instead, it flatly contradicted the whole thing.

When we discovered this deflating state of affairs Smithy and I escaped from the threatening press interviews to a secluded office at the airport, and tried to work out what we could do about it. On present evidence we couldn't fly the Pacific at all, because even with the airplane tanked up to double its allowed capacity for the air race and being in fact a flying fuel-tank system, it still hadn't enough fuel to reach Honolulu at the recommended power settings.

We decided to do some tests in the air, on two-way flights over measured distances, at a series of different power settings—and see what results we could get.

It worked.

The drama was over.

We finally came in with a series of weight, power-setting, and consumption combinations which could put us in to Honolulu with a two-hour fuel margin against a 20-knot head wind. That was a reasonable flight condition.

Immediately before I closed the perspex hood over my cockpit for departure from Brisbane the next morning a woman rushed out of the crowd and handed me a beautiful white rose. I had never seen her before, but the spontaneous gesture and her words "Wear this; for luck" struck a good note on departure. I waved to her, and drew the stalk of the rose into the buttonhole of my coat. All the seven thousand miles across the Pacific I protected that rose and as long as it was there I felt that our engine would keep going.

We flew through that day to Suva in the Fiji Islands. After an "on top" crossing of the Coral Sea above big cumulus build-ups we came in over the brilliant colors of the protecting reef off New Caledonia, located the aircraft over the capital, Nouméa, and then set course for Suva.

East of New Caledonia we ran into weather, dull nimbus cloud with rain, with cumulus below us, and threatening conditions. Later in the afternoon I got brief sights of the sun astern and was able to fix our distance run, but with the sun out of position to check the track, only accurate dead-reckoning could keep us for Suva, and in the cloud and rain I simply had to guess rather than use the tools of my trade. Eventually we descended and came out the bottom of congested weather quite close to the sea, over a leaden, menacing ocean upon which rain fell with black squalls of wind.

There is a temptation in such conditions, with a critical island objective, to go hunting in the shadows for land, to abandon the course set upon the compass from the best information one can muster in the circumstances. But in this temptation there is confusion which has to be resisted and rejected, and the aircraft kept on a course which is the result of sound reasoning. Action upon wishful thinking for the sight of land is likely to lead to disaster.

Smithy and I sat silently in our cockpits, with a final course on the compass and, with the setting of the sun, the light ahead in the east fading and already showing signs of a blue-gray darkness. It was beginning to look as though we should have to pull upon the climb and try to break out through the top for stars to give us a position for a night descent and approach: an unattractive prospect.

And then we saw it. The unmistakable sign of the coast; a line of surf breaking white on the south coast of Viti Levu.

We passed in, by the entrance to the Singatoke River and on, low over the reefs, to Suva. In the very last of the light Kingsford Smith put the airplane down in Albert Park, a cricket field in the town with an available run of about 300 yards.

We flew out of Albert Park with nothing in the airplane except a few gallons of fuel to take us the twenty miles to Naselai Beach, our take-off run for Honolulu. Strong cross winds held us down at Naselai for a week, but early in the morning of October 29 the air was still, and with a low tide exposing a fine stretch of hard gray sand ahead of us the Altair, overloaded with every fuel tank full to the filler caps, was airborne in less than a mile.

Soon the Fiji Islands passed away astern and we headed out into the great open spaces of the Pacific. The aircraft had by this time so impressed itself upon us that we had no thought of engine failure, and our minds were fully occupied with wind and weather tactics, control of the aircraft, and navigation. In the afternoon we came up with the Phoenix Islands and the sight of these lonely atolls only impressed upon us the vastness of the ocean and the space above us. But we saw their material value as a refueling base for aircraft flying south from Honolulu.

The steady drumming of the motor and the spread of the Altair wing carried us on into the night over the scattered cumulus cloud of the Pacific. But early in the night the stars were blacked out ahead and we ran into weather. After half an hour in heavy rain and turbulence that snatched and flung the aircraft in a turmoil of unstable air, one of those freak incidents occurred that would probably not happen again in many flying lives.

Kingsford Smith was switching on the landing lights occasionally to see if the rain was easing up at all, because we were a little concerned about the effect of very heavy rain, since, on the flight to Suva, some damage had been done, by rain, to the leading edge of the wing. We had fabricked over the damage, but we didn't like the sort of violent torrent of rain through which the aircraft was now slashing her way in jet black darkness.

Quite suddenly there was a general lessening of sound and it seemed that we were slowing up. I looked down to the airspeed indicator and found to my astonishment that it was indicating only 90 knots, after the steady 125 indicated airspeed at which we had been cruising. We were near to stalling speed and it was obvious that Kingsford Smith was having difficulty in controlling the airplane. I looked at the duplicate throttle control in my cockpit and was alarmed to see it right forward, for all the power the engine would give. Something was seriously wrong: yet the engine was running without any obvious signs of failure, but with an impression of laboring, at full throttle and only 90 knots on the A.S.I.

Then I felt her go . . . suddenly, into a spin. The night became a whirling madness, with the needle of the turn indicator hard over, the rate-of-climb indicator showing a rapid descent, and the needle of the sensitive altimeter winding its way down the height scale. From fifteen thousand feet, where we had tried unsuccessfully to climb over the weather, the height was going rapidly and the airplane was fixed in the sinister rhythm of the spin.

Kingsford Smith came in on the intercom, "I'm sorry, but I can't get her out."

The tone of this remark was one of regret, almost of apology, that I was going to be spun into the sea. It rocked me considerably. I called through to Smithy, "Do you mind if I have a go at her?"

"Yes; go ahead," the answer came back quite calmly.

I took the controls and immediately took the emergency action for recovery from a spin. Nothing happened. The airplane went on spinning in the whirling madness of the night. I held the stick fully forward and planted my foot firmly on the opposite rudder to the direction of spin, but it seemed that I had lost all contact with the Altair, that the familiar sense of complete control was lost entirely and I was left pinned in the seat with no effective approach to the airplane.

To the dramatic effects of the spin was added the shrieking of the klaxon which on this airplane was set to sound off when the throttle was closed with the undercarriage up, a safety provision against the pilot forgetting to have the gear down and locked for landing.

It was not a time for prolonged conference about the situation. From fifteen thousand feet the altimeter now was down to six thousand. Since he was the pilot of the airplane Kingsford Smith, by mutual consent in the briefest of conversations, took over again and I deliberately took my hand and feet off the controls, which I saw immediately being moved in decisive fashion to the limits of their travel in various directions. My natural instinct as a pilot was to go on trying to get out of the spin, and it was a terrible anticlimax to just abandon the whole thing, and Smithy was about the only pilot I knew to whom I could have handed back the airplane at this stage.

Then something was changing in the behavior of the aircraft. The whirling feeling had gone and we seemed to be diving vertically for the ocean. My eyes flicked to the instruments. The turn indicator was steady but the altimeter was still winding off height. Then I felt myself being pressed into the seat, the rate of descent was easing up, and the airspeed coming back.

She was out of the spin . . . and coming up out of a dive.

Very soon he had her leveled off, and flying. But the airspeed was back in the region of 90 knots and the throttle lever fully forward again.

Down in the lower levels the turbulence was less violent, and Smithy had the airplane under control; but still flying in this unaccountable fashion, at full throttle and 90 knots, with a kind of rumbling stress in the engine and a slow lifelessness in the flight of the aircraft. Then, to my intense relief, I felt life returning to the Altair. She was starting to flow again; to bore on through the ocean of coal-black air with alertness and decision. At the same time Smithy's voice came through again, "I've got it . . . the flaps. The flap switch was down. I must have knocked it on when I was using the landing-light switch to check the rain."

So that was it; obviously; for now the aircraft was indicating her normal cruising speed, flying cleanly, and the throttle back to the regular cruise setting.

I began to think back, first with relief from the depressing deductions of remaining range if we had had to go on flying at 90 knots and full throttle. It is always this way in the air. Immediately when a crisis occurs, one takes all possible emergency action to eliminate the cause and at the same time quite instinctively works out the effects and the action to be taken if it cannot be eliminated. As soon as we were out of the spin and there was reason to believe that the airplane could go on flying, even though precariously, the briefest calculation had put Honolulu out of the picture. Using all that power with such poor airspeed we should have had dry tanks hundreds of miles before the Hawaiian Islands: so our thoughts had gone to Fanning Island, a relatively small atoll about five hundred miles from our estimated position. The thought of finding Fanning Island in the darkness under the chaotic conditions in which we still flew, even though under control, was somewhat theoretical; but I had found myself quite systematically sorting out a plan of navigation for Fanning Island. It was then that Smithy discovered the flap switch, selected Flaps Up, and away we went on course for Honolulu.

And, the aircraft's behavior had been quite reasonable. Slowed up by the terrific drag of full flaps, we had had little margin above stalling speed to cover the inevitable fluctuations in the violent turbulence of what was obviously the bad, higher region of the cumulonimbus cloud of the intertropic front, particularly active this night.

With full flaps down, the normal airstream over the tail control surfaces had been deflected in the spin and was apparently not direct or strong enough to bring about recovery by normal control reactions. Something in the odd movements of the controls to which Smithy finally resorted as we spun down for the sea must have upset the rhythm of the spin and converted it into a clean dive, from which he was able to pull her out, on instruments.

As so often happens with our life in the air, soon after the crisis had passed, the scene changed dramatically and we broke out into a perfect clear night with tops of broken cloud below us at about seven thousand feet, and all the stars in the heavens brilliantly welcoming us on into the North Pacific. Polaris, the North Star, was there ahead off the port bow in the northern sky, and Sirius, the brightest of all the "fixed" stars, lay near enough abeam to check our track.

Order was coming back into the night. I had retrieved my scattered navigation tables, workbook and instruments, the sextant and chronometer having remained secured against violent movement of the aircraft. Since we were now moving into northern latitudes I took a bearing on Polaris with a special bearing plate I had devised and fitted on the deck above, under the perspex, and from this bearing off the ship's head, related to variation and the compass course, found the new deviation of the compass.

Altitudes of Polaris and of Sirius with the sextant gave us

our position, and the whole situation was tidied up. We set course now from this known position in the wide spaces of the starlit Pacific night, and a wonderful peace settled upon us. In these ideal conditions I stepped up our latitude each half hour from Polaris, and to check the track of the Altair took whichever star was conveniently abeam, for a position line. She was boring cleanly ahead for our objective in the North Pacific, with little correction to the compass course.

As dawn came we looked into the northeast for a sight of the 13,000-foot twin mountains of the big island of Hawaii; but these were not visible in the haze and there were only sky, cloud, and sea in this morning of approach to our objective. At this time of year, the end of October, the declination of the sun (its latitude in the heavens) south of the equator brought it almost dead abeam of our track, so I called through to Smithy when we were still some three hundred estimated miles from Honolulu.

"How about letting down now? I'd like to go down right to sea level, set the altimeter; then level off at 225 feet and use the natural horizon to get a good position line for the run in. Two twenty-five will cancel out the allowances and give us an accurate altitude right off the sextant."

"Right. All set to go down now?"

"Yes. The usual rate of descent should do it."

From eight thousand feet we were at sea level in twenty minutes, skating almost on the surface. It was fine to have the sensation of speed and visible progress about ten feet from the surface. I set the altimeter, called through to Smithy, and in less than a minute we were leveled off at 225 feet. The air was quite turbulent under the broken cloud, but I managed to lay the sun clearly on the horizon as we passed through one of the sunlit patches; took the chronometer time, closed the perspex hatch, and set out books, pencils, dividers and all personal gadgets, to work and lay down this vitally important position line.

It put us eight miles east of the track. I passed the course alteration forward with a good, neat feeling of finality. Then, to be sure, I slid open the hatch, took another sight, and worked it. The position line tallied exactly with the first. I put away all my gadgets, tables, and equipment and sat back to relax and wait for Honolulu.

Fuel? How were we going for fuel remaining? We had checked on time in air, power settings, and consumptions, and it looked good. We had no fuel gauges, but reckoning up again from our own test figures it looked like two hours' fuel remaining at Honolulu. Smithy had run out the big fuselage tank, all the wing tanks, and recently turned over on to the last tank, the ninety-gallon gravity between our two cockpits.

The relief of a well fixed position for approach having passed, we now started that familiar straining into the distance ahead for the first sight of land.

Smithy was the first to see it, through his screen ahead: a thin gray outline which I picked up as he turned the aircraft off course to bring it into my vision. There was no doubt about it. This was land. The incredibly satisfying land of the Hawaiian Islands; land which only now proved to be real, after the infinity of day and night in the depths of Pacific space.

I just sat there, filled with a curious sense of gratitude that we had been given the conditions to find these islands, that the engine had never shown a sign of failure in the 25 hours of flight; and the most wonderful sense of anticipation for arrival at Honolulu.

As we approached the island of Oahu over the deep blue of the Pacific we could see the shallowing coral shore, bright with the lightened colors of the sea; and the buildings of Honolulu and Waikiki lying white and colorful in the sun.

We cut in to the shore and flew over Honolulu, and then turned away by Pearl Harbor for Wheeler Field. A formation of United States aircraft came out to escort us in and soon we could see the hangars of the Air Corps station. The everlasting sound of the engine subsides to a rumbling purr. Gear down.

She came away in a turn for final approach.

Flaps . . . and the cool, green surface of the airfield close below. There was a strange silence as the Altair floated the last few yards; then the wheels took the load as she lost flying speed. We taxied over the grass, to the concrete apron in front of the hangar. An American officer signaled us in, and she came to rest. The engine rumbled over the last few compressions, and stopped.

Silence.

For a few moments there seemed to be a complete cessation of life. I remained still in the cockpit, without thought or movement. Then I ran back the cabin top and looked about us. There was a huge crowd of people along the tarmac, and a group already surrounding the airplane. We were quickly overwhelmed by the enthusiasm of our reception.

Still far away from the earth, and belonging still in the cockpit of the Altair with stars and the infinity of night over the Pacific, I was not yet ready for these warmhearted people. Shaken by the hand from all sides, the perfume of the leis hung around my neck in ever mounting volume slowly established my first real contact with the island. I suddenly thought of my white rose, and found it still in the lapel of my coat.

As we passed through the first overwhelming flood of our first reception we met Major General Halsted Dorey, commanding officer of the Hawaiian Station, and other leaders of the community, and then we passed on to the only formality of our arrival, clearance through customs. Somewhere in the archives at Honolulu today is an interesting document, the clearance papers for the first international aircraft ever to pass through customs in Hawaii, the Australian-registered Altair, VH-USB, on October 29, 1934.

But there was a sequel to our arrival that day at Honolulu, an event in the true dramatic style that could hardly have been more effective in sustaining the sensational news of our flight; although, without a fully published explanation of its cause, it was not good in our view as professional aviators.

After a triumphal entry to Honolulu in gigantic motorcars escorted by the traditional motorcycle police, the great wave of hospitality surged on to the Royal Hawaiian Hotel where we were entertained as guests in the most lavish fashion with a large and glamorous suite facing out over the surf at Waikiki, to the South Pacific of our flight.

The luxury of all this was so satisfying that it would be futile to try to describe it. The contrast of luxuriating in a clean bath, eating the perfect pineapple of Hawaii, attended with the warmest hospitality, after the experience of spinning down into the night with the klaxon screaming and the invisible ocean coming up with the most uncompromising climax, was really something to be lived but not to be embellished with attempts at description.

So, under the influence of all this, and with the most benevolent outlook to the whole world, Kingsford Smith, though he had had no sleep for more than thirty hours, invited the mayor of Honolulu to a flight over his city in the afternoon.

I went out with them and watched the Altair take off and head east for Honolulu. They had been airborne for about three minutes and were at about two thousand feet when I saw the aircraft come away in a descending turn, back towards Wheeler Field. As they came closer, still descending, I saw that the propeller was just whistling around, windmilling without power.

Without any attempt to circuit the airport to line up for a landing into wind Smithy came straight on in, still with the gear up and obviously stretching the glide as far as possible. Then, as he became sure of making it downwind, the undercarriage came down; then the flaps as he came over the boundary, and the Altair settled quite comfortably onto the grass, ran some distance, and came to a halt with the propeller stopped.

I went out in the car to link up with Smithy in the airplane. The engine had just stopped, after a few spluttering coughs; like a fuel stoppage.

The lines were checked. No fuel coming through.

Then the tank was dipped.

No fuel.

The entire fuel system of this flying tanker was empty. It was simple. They were out of fuel, in three minutes' flying after we had landed, from Fiji.

The significance of this situation began to creep up my spine. Notwithstanding our careful fuel check flights at Brisbane, estimating two hours' fuel remaining at Honolulu, we had landed with about five minutes' petrol. But it had been painless, for we had known nothing of it coming in from 25 hours' flight over the ocean. Still the whole matter had to be thoroughly investigated before we set out over the 2090 nautical miles of ocean to San Francisco.

The Altair was hauled into the hangar and our friends of the (then) U.S. Army Air Corps went to work on her. The fuel system was dismantled and the fuselage tanks taken out.

It wasn't long before the mystery was solved.

There was a large crack in the ninety-gallon gravity tank in the fuselage which had let about two hours' fuel leak out and drain away through the bottom of the airplane. Our tests and consumption calculations had been proven right, but a really bad situation was discovered in the bottom of the big, main fuselage tank. A too prominent rivet head in one of the tank bearers had rubbed through the felt packing between the bearer and the tank and had gone on rubbing, under normal vibration, through the tank itself until there remained a layer about as thick as paper! We had to have this whole tank and fuel system out of the airplane, inspected, and put back so that it wouldn't leak. After four days and nights of Hawaii we were ready for the flight on to California.

The Altair took us across the North Pacific to San Francisco in fifteen hours of steady flight, through a night that was easy for navigation, over cloud that covered the darkness of the ocean. Polaris, the North Star, was prominent again in the stars we used for navigation; and in the dawn the tops of the San Francisco hills were clear ahead over a layer of surface cloud which completely obscured Oakland Airport.

With a good deal of fuel reserve, we floated around at low power, sorting out the tops of the hills to locate the position of the airport, and it was not long before the fine surface cloud began to clear under the influence of the sun: and there below us our final destination came into our view from the quiet rumbling of the Altair sailing smoothly on the morning air. In a few minutes we were on the ground, and stopped, by the airport buildings.

This time there was no moment for reflection before the human storm of excitement burst over us. Newsmen jumped up on the wing with microphones to record the first words of the crew of this first aircraft ever to arrive across the Pacific from Australia. Others with note pads flourished pencils seeking the story for their newspapers; and around all this was a seething crowd of people waving and calling to us, many with pieces of paper to be autographed.

The tempo was different from Hawaii. Here we had plunged a story into the already high frequency of life in the United States, and the reaction was terrific, still with the same good humored, really personal note of reception as at Honolulu, but stepped up to a higher tempo of excitement which made me wonder what would happen when we got out of the airplane, now under guard by police. Of all the people the one who most caught my attention was a small American boy, his eyes blazing with excitement, who succeeded in maneuvering his way to the leading edge of the wing and reaching us. The spirit of adventure was surging within him. I passed him my rose, told him its story, and told *him* to keep it, for luck.*

---

* As a consequence of this flight the lonely Phoenix Islands stood in my mind as the obvious refueling base for aircraft flying south from Honolulu. By tanking up the Altair far beyond her normal gross weight we had been able to make the long flight from Fiji to Hawaii; but aircraft of the future, particularly those operating a regular service with passengers, would not, for many years, be able to carry enough fuel for such a distance; and alternate airports also would be needed between Honolulu and Suva for safety in the case of emergency landings. So when I returned to Australia I saw the Minister for Defence (then the relevant Minister) and suggested that we should make a move to establish a base airport in the Phoenix Islands. I was totally unable to initiate any immediate action from Australia; and it was made very clear that I was regarded as some sort of visionary and highly inconvenient phenomenon. Finally, after no little time and no less persistence, a party was sent to Canton Island from New Zealand, the first to occupy the lonely ring of coral sand and rock which, a few years later as an airport, was to be the vital refueling base enabling United States military aircraft to reach Australia from Honolulu and, in service with the allied air forces, help destroy the invading Japanese fleets.

Chapter 7
Jubilee Mail, 1935

VERY SOON after I returned to Australia a proposal came up for another trans-Tasman flight, this time with two aircraft carrying a special mail, to commemorate the Jubilee of Their Majesties King George V and Queen Mary, in May of 1935. Jack Percival, who had been in the *Southern Cross* on the Gerringong Beach—New Plymouth flight in January 1933 (when I was learning to navigate the airplane) conceived and planned this Jubilee Mail flight, which was designed also to create further public interest in the inauguration of a regular mail and passenger service between Australia and New Zealand.

It was intended to be the last trans-Tasman flight of the gallant but aging *Southern Cross*—and it was.

Kingsford Smith, of course, was taking the *Cross;* and I was invited to go in command of the second aircraft, *Faith in Australia,* Charles Ulm's modified Fokker, in which he also had made a number of pioneer Tasman crossings and other

59

flights. The trustees of his Estate had made the aircraft available for this special commemorative mail.*

On the day of departure we flew both aircraft to the Royal Australian Air Force Airdrome at Richmond for night take-offs, estimating daylight landings at New Plymouth.

Soon after our arrival at Richmond it was revealed that Kingsford Smith's navigator was ill and could not go on the flight. A period of high drama, typical of such situations, prevailed for several hours. Eventually it was decided to take only the *Southern Cross*; with Smithy as commander, John Stannage as radio officer, and myself as navigator and relief pilot.

I could hardly have been more apprehensive about this turn of events. While supervising the work on my own aircraft, I could not help noticing that the day before departure one of the engines of the *Southern Cross,* lying dismantled in pieces on the hangar floor, was being assembled by John Stannage, an incredibly good radio operator and technician, and by Jack Percival, a first-class official correspondent on the flight, and the man who had conceived and very efficiently organized the whole project; but neither was an engineer. Kingsford Smith had a way of making such situations work out perfectly well in practice, but from the moment I realized I was not taking Ulm's aircraft and was committed to travel in an airplane one engine of which had been assembled in circumstances which absolutely horrified me, I could see little future in the whole thing.

I was a little touchy about situations like this because, not very long ago I had escaped from another unpromising affair,

---

* Just days after the arrival of Kingsford Smith and myself in Oakland in the Altair, Charles Ulm and his crew, Littlejohn and Skilling, had set out from San Francisco in the small, twin-engined Airspeed Envoy aircraft, *Stella Australis,* bound for Honolulu and Australia in an attempt to make the second westbound crossing of the Pacific. At the end of his fuel, hopelessly lost in bad weather but still searching for the Hawaiian Islands, Ulm had sent out his last and typically laconic message: "We are now landing in the sea. Please come and pick us out." Tragically—despite a massive air and sea search—they were not found. Theirs was a great loss, personally to us all and in the case of Charles Ulm, it removed from the scene a man who, had he lived on, would have been at the top of international air transport today.

on the very brink of what was intended to be a transatlantic
flight.

I was navigator of Ulm's *Faith in Australia* on an attempted round-the-world flight which, after various structural
failures in the engines, had reached Ireland, westbound from
Australia.

The airplane was standing on Portmarnock Beach being
fueled to full tanks for the Atlantic crossing. On the record
of the engines from Australia such a crossing was quite theoretical, and in any realistic view of an attempt to make it
there were few redeeming features.

I was in the airplane, in that mental state of acceptance
one has to develop in wars for psychological survival, and
holding the fuel hose to top up the last few gallons of the last
cabin tank when, with a fearful scrunching noise, the undercarriage collapsed and the airplane fell down on the sand. I
could not have been more relieved, because here was an escape, at least temporarily, from engine failure and the Atlantic.

By further good fortune, the Atlantic westerlies had set in
by the time the airplane had been repaired, and we couldn't
make the westbound crossing. Instead, after complete overhaul of the engines, we flew continuously, except for replac-
ing some pistons at Calcutta and being bogged in soft ground
at Surabaya, for six days and nineteen hours eastwards, thus
making a record flight back to Australia; cruising speed 80
knots.

But, to return to the *Southern Cross*, I unloaded all my
gear from the *Faith in Australia*, set myself up in business in
the airplane, and was as ready to go as a navigator with any
imagination could be in such circumstances. But I had to
admit to myself as we prepared for take-off that there was
something about this airplane—something good inhabiting it
—which made me feel that, for no reason I could put my
finger on, the *Southern Cross* would not fail us.

As midnight approached Kingsford Smith started the motors. I took my take-off position in the starboard pilot's seat
and listened to the tearing snarl as each engine ran up to full
throttle, and their shattering blast came through the open
sides of the cockpit. Very heavily overloaded with fuel, and

now with all the mail, and some freight, she taxied slowly out for the take-off; and into position for the longest run on the airfield. There she faced the night with a steady, bellowing roar and slowly moved away.

There was the familiar thunderous stress as she fought her way to speed for flight, and near the end of the airfield the change came with relief, from earth to air; from all doubts and confusion, to an aircraft, airborne and passing into the quiet intimacy of the night where the sound of the motors and the airstream becomes an unnoticed accompaniment to living.

The night was clear and bright as the *Southern Cross* moved across the light-studded land north of Sydney with a steady purpose in her flight. Soon the coast of Australia came in below and we passed out into the Tasman night. I went below to take back bearings for departure on Norah Head and Macquarie Light. Both stayed bright on the horizon till we were far out from the land, but an hour from Richmond the last flicker disappeared with the world we had left.

The *Cross* was alone, a thing apart from land or sea, steady and sure in space, having no connection in my mind with an aircraft one of whose engines had been strewn in pieces on the floor of a hangar only a few hours ago.

A hundred miles out we ran under a layer of scattered cloud which built up as we flew into the east. As this suggested some southerly weather I went below to let go a flare and check the drift.

The first was a dud. No light showed upon the sea. I let go another, and waited for the point of light to show in the darkness down behind the tail. Far back in the night it seemed to leap up out of the sea in flame; then fade to a glowing point of light moving away astern and to the south.

I reckoned eight degrees of port drift, gave John Stannage the course and dead-reckoned position for transmission, and went forward to give Kingsford Smith the new course to steer. The cloud had shut in to scattered showers of rain and Smithy was flying her on instruments, holding three thousand feet of height above the sea. It was too soon yet with the still heavy overload to think of making height for the westerly.

At about five o'clock I took over to give him a spell from

the flying and he went below to see about some wireless messages to Sydney. Between the blind regions of the rain showers it was just possible now to see a faint horizon over the nose of the aircraft, and from the pilot's seat I could see the flame-heated exhaust manifold glowing brightly out over the center motor.

Lifting my eyes occasionally from the flight instruments to take in the early morning weather as signs of light came into the east, I saw nothing unusual in the red glow of the exhaust ring. All my senses were in harmony with the sound, the sight, and the touch of the aircraft and the air, and I sat relaxed and happy, flying into the dawn.

But suddenly I was alerted to a change. Just one small spot on top of the exhaust manifold on the starboard side of the center motor was glowing with a lighter, brighter color than all the other visible parts of the exhaust ring. I looked quickly to the manifolds on the outer motors. The glow was steady and clean, with no light spots on the metal. With all the warning signals up, I flew the aircraft instinctively, concentrating on the exhaust of the center engine. The unusual light was there, and could not be denied. But since nothing could be done about it I kept a close watch on it and began to take in the now visible weather effects upon the navigation. I wanted to pick up the wind force and direction from the appearance of the sea, since there would be little, if any, variation at our low altitude, below the cloud base.

As the light increased, the surface of the sea showed a strong breeze from a little west of south, almost dead abeam. I signaled back to Smithy that I needed to go aft for a drift sight. But at the same moment the importance of any normal working of the aircraft was canceled by unmistakable signs on the manifold. The welded edge of the pipe had split, and through it the exhaust was blowing in a flickering slit of light from the trailing edge. Even as I watched, the blow of the flaming exhaust was gradually forcing open the crack and bursting open the whole top of the manifold.

At that moment Kingsford Smith returned to the cockpit and took over so that I could go aft for the drift sight. But when he was settled at the controls I drew his attention to the state of the center manifold. We both sat fascinated but without comment, watching the rapidly disintegrating pipe, till in a few moments the whole top section was blasted out by the flame, flicked away in the airstream and was gone.

Instantly the most terrific vibration shook the aircraft as though some giant, invisible hand had reached out to shake the life out of her. Mentally, my hand flew to the throttles, but Smithy was flying the *Cross* and his sure hand was there. He drew off the starboard throttle and we both looked out to the motor. It leapt and struggled in its mounting as though it had gone mad and was trying to wrench itself out of the airplane.

Through the fuselage a sickly, pulsating wobble shook the *Southern Cross* as the slowing propeller lashed the air: and as we finally saw the blades and they came to rest, one stuck out towards us in broken, splintered wood; a jagged stump, like a lightning-struck tree.

Smithy held up the *Cross* with two engines at full throttle, but she started to sink towards the sea. A few words passed between us and he turned her away and headed her back for Australia. As an approximate course I clapped 285 degrees on the compass to keep the wind no worse than abeam and to give us the best speed towards the nearest land. It seemed quite theoretical, to be heading for land more than five hundred miles away when at full throttle the altimeter needle was steadily sinking down from the level of three thousand feet.

Weight. That was the thing. Somehow we would have to get rid of weight. Smithy was fully occupied holding the *Cross* up to the best attitude for flight, and was holding every possible inch of the falling height; but we were obviously destined for the sea within less than half an hour. I shouted across to Smithy, "Have to dump some weight. Shall I go ahead?"

His voice came back in the snarling roar of the extended motors, "Anything except the mail."

I slipped below to the cabin, passed the word to Stannage to dump everything except the mail, and then turned on the dump valve of the main cabin fuel tank. How much to dump? That would have to be worked out immediately before too much drained away.

We had been in the air nearly seven hours. Say seven

hours at thirty gallons an hour; 210 gallons gone; 390 gallons left.

I went to the chart and estimated our position and distance out; from Australia—590 miles. Nearly half the distance to New Zealand: but best to go for Australia. Weather and head winds the New Zealand end. Say, six hundred miles to the Australian coast. Speed, with the nearly stalling aircraft, about sixty-five. Wind abeam. Make good her airspeed. Reckon it at sixty. Six hundred miles at sixty. Ten hours.

Ten hours on two motors! Best not to think too much about that. I remembered the rate of flow of the dump valve, and turned off the cock till I got it all sorted out. A glance up into the cockpit to the altimeter. About two thousand feet now.

Ten hours at 28 gallons an hour on two engines. She'd use that, taking out all that power: 280 gallons. Say three hundred. We must keep at least three hundred gallons.

It may appear very risky to have left only enough fuel to reach the coast with so narrow a margin, but this was a risk which had to be accepted against the certainty of descent into the sea. I knew the aircraft would sink within a few minutes. We had no dinghy; nor even life jackets, in the *Cross*. So the picture was clear. The mail had to be kept until the very last emergency. So we had to dump the fuel.

I reckoned up the amount in the top tanks, unscrewed the filler cap of the cabin tank and dipped it with the measuring stick. We could let go more fuel. So I turned on the dump valve again and kept a watch on the decreasing level, with the dip stick.

Finally, leaving a little more than the total of three hundred gallons, I turned off the valve and checked the altimeter. She was down to five hundred feet now, but holding the height: so I left it at that. The few extra gallons would not put her in the sea now. Luggage, tools, freight, and all articles not essential to flight had gone out into the Tasman Sea. Only the mail remained; the bags lashed down in the cabin behind the big tank.

I went up front, to tell Smithy about the fuel, and to let him know everything had gone overboard.

There, it was as I had expected. He was settled down, but

extended; holding the *Cross* in the air; and his aircraft, feeling the master touch, leaned heavily on the air, staggering; but flying. He held her with the wheel, feeling just where her strength lay; using that, and not overburdening her weakness. He felt her through his hands and feet, and the seat in which he sat, trying for support from the slowed-up airstream: and he lay her wing upon it at exactly the right angle, the only angle, at which she could fly and maintain height.

Down in the cabin again, I went back to John Stannage and his radio. We exchanged a smile of appreciation. We found some humor now in the fact that we were not immediately going down in the sea. This reprieve brought with it a delicious lightheartedness that was in strong contrast to the threatened disintegration of our world only a short time ago. The airplane now was not shaking itself to pieces; it was not losing height; and that was enough. We really felt quite lighthearted, and did not yet choose to look into the future at all.

Stannage had been in contact with Sydney; reported the broken propeller and the precarious situation of the aircraft; and had given our position, course, and speed. Our clear objective now was to reach land: not Sydney airport, but Australia. The nearest land was at Port Stephens, where the coast bends out to the northeast at Stephens Point. There was little difference in the distance; but by laying off north of the track to Sydney we could bring the wind more abeam and make a better speed. I gave Smithy a compass course for Seal Rocks, 120 miles north of Sydney, and when he straightened the *Cross* up on this course the wind was slightly better than abeam.

Up there in the cockpit the two throttle levers were still right forward, taking all the power the two remaining engines could give. There was a drastic finality about the sight of those throttle levers, proclaiming the fact that we had no reserve and were just maintaining height at three hundred feet. But the old motors of the *Cross* were snarling defiance at the ocean in the harsh, blaring crackle of their exhausts. We were afloat in the air, even though precariously, and flying; and we did not think too much about how long the engines would keep going, dragging a dead motor and propeller on the starboard side, a still heavy load, and a wing obliged to meet the air at an attitude of great resistance to fly at all. But we hoped they would last till the reduction of weight as they burned down the fuel would allow us to ease them down from continuous maximum power.

As we made some distance westward the showers of rain passed, and through the broken cloud shafts of sunlight brought life to the dull gray world of the ocean. The sun was nearly abeam to the north on a bearing suitable for a position line to check the track of the aircraft. There was too much turbulence for accurate results with the bubble sextant; so, to give me the natural sea horizon, Smithy eased the *Cross* down to a few feet above the sea and I was able to get a good set of sights. Worked, and laid down on the chart, the resulting position line showed us to be making good the track for Seal Rocks.

Over the radio from Sydney we learned now of the action being taken for our rescue. The pilot vessel, *Captain Cook*, had left to intercept our track; H.M.S. *Sussex* would be under way in three hours; and *Faith in Australia* would leave as soon as a suitable pilot could be found for her. All this warmed our hearts considerably and was in principle very reassuring, but to stay in the air and reach land was not only the clear objective for survival, but we were now to have ambitions for return to Sydney airport and a normal landing. It was not long before we were back on the single objective of survival, for the aircraft and ourselves.

For some time I had noticed a steady stream of blue smoke in the exhaust of the port engine. There wasn't much; but it was there, coming away in a continuous streak and very visible in the clear air. It was obvious that this engine was burning oil. There were no quantity gauges on the oil tanks, each situated inside the cowling behind its engine, and therefore no way to measure the amount of oil remaining in the tank. It was assessed from the known consumption of the engine, and normally there was a big margin of oil beyond the range of fuel. Each tank held eleven gallons of oil and normal consumption was less than a quart an hour. Now, with the evidence of this ominous blue stream from the port exhaust, my imagination saw right into a tank with not enough oil to reach Australia. Suppose the engine was burning

a gallon an hour. An old engine, wide in the clearances,
being thrashed to death at maximum power: it could be
burning a gallon an hour; and we had been in the air now for
nearly eleven hours. Even allowing for more normal consumption over the first seven hours, at high cruising power,
the outlook was not good.

I thought around this problem a good deal, and it kept
coming back at me. Eventually I tried to accept this blue
smoke and hope that I was wrong about the consumption;
but the oil pressure gauge of the port engine now had a fatal
fascination for me, and my eyes were never long away from
it. I said nothing about it to Smithy or John, because talk
could not improve the situation, and in the remote vent that
they had not noticed it there was no point in passing on such
depressing possibilities in a situation already loaded with sinister implications. But the confidence and relaxation which I
was beginning to experience as the *Cross* continued to stay in
the air and put more of the Tasman Sea behind her were
completely ruined by this infernal blue stream of oil smoke,
since even the most optimistic wishful thinking could not
admit the remotest possibility of the aircraft remaining in the
air on one engine. The sea was again the final abyss, and the
Cross our world hanging precariously above it.

Earlier in the situation I had attempted to cut off the ends
of the starboard propeller blades with a hacksaw. I thought
that if I could trim off the shattered blade, and cut the other
to the same length, we could at least let this propeller wind-
mill, and might even get some thrust from it using some
throttle with the engine.

One of Smithy's problems in flying the aircraft was to prevent the airstream turning the broken propeller; for, immediately it started to turn, the unbalanced forces of the blades
set up the most appalling vibration which soon would have
started the disintegration of the aircraft. Any increase in airspeed above the absolute minimum for flight would set this
propeller windmilling and Smithy would have to haul the
Cross up almost to stalling speed to stop it, and then very
carefully ease her down again to the very narrow margin between stalling and windmilling the propeller. This was a ter-

rific strain for a pilot and I had tried to eliminate it by trimming the blades to a more balanced condition.

To attempt this operation I had gone partly out into the airstream from the open side of the pilot's cabin; but the blast of air, and the fact that the propeller would turn every time I tried to work on it with the hacksaw, finally convinced me that there was no future in this idea, and I just slumped back into the cabin, exhausted and frustrated.

But now, with the evidence of the blue smoke trail continuously before me, I began again to think of some way to improve our situation. It was quite uncomplicated, really. If the port motor used all its oil the engine would be destroyed. With the center motor alone we would be in the sea within a few minutes. There the aircraft would sink, and if we happened to survive the ditching with a field undercarriage aircraft, we would stay afloat just as long as we could go on swimming in a rough sea without life jackets. There was a strong incentive to do something about oil for the port engine.

I began to speculate about the possibility of somehow getting oil from the tank in the cowl behind the useless starboard engine. There should be at least nine gallons of oil there. If some way could be devised to get this oil, and somehow transfer it to the tank of the port engine, we should have enough oil to keep the port motor going to reach the coast.

Every way I looked at it there was obviously no straightforward way to make this oil transfer, since each engine was a complete unit of its own, with no lines or pipes interconnected. The outboard engines were isolated alone, far out in the airstream under the wing.

After developing every line of thought without any tangible result, it wasn't long before I reached the alarming conclusion that the only way to do this oil transfer was to go out and get the oil from the starboard side and go out again to put it into the tank on the port side. With the results of the propeller-trimming episode fresh in my mind this final conclusion was a very unattractive prospect, but rather than live with defeat in my mind, and with what I now believed was the certainty of being forced down in the ocean, I let this idea of going out in the airstream to the engines support my morale, which was in need of some hopeful outlook at this time. As the idea gained some momentum I found myself starting to work out the details of some practical plan. In the beginning it seemed entirely theoretical, like thinking of flying to the moon (not so theoretical now); but as the plan developed in my mind it began to seem less impossible, and as we flew on low over the ocean I began to see it as something which was at least positive thinking, which freed me from a dumb acceptance of ending up in the bleak and threatening Tasman Sea.

The outboard engine nacelle could not be reached directly from the open side of the pilot's cabin; but out from the fuselage below this window a streamlined horizontal steel tube extended to the frame of the engine mounting. It was part of the lateral bracing system for the engine and the undercarriage leg, and was quite strong. I wondered whether I could get out the side window of the pilot's cabin, stand on this strut in the airstream with my shoulders against the leading edge of the wing, and somehow move out sideways and reach the engine. If I could do that, and hold on out there, I could unclip the side cowl, perhaps reach the drain plug of the oil tank, undo it, and drain out some oil in some sort of container. Then, if I could get back along the strut and into the cabin again, it would mean going out the other side, unscrewing the oil tank filler cap, and pouring in the oil I had collected from the starboard tank. Apparent impossibilities came back at me from this plan—the force of the slipstream, the precariousness of trying to stand on the strut, how could I collect the oil while somehow holding on out in the blast of air? How could I get back with the oil? Then there was the other side.

Impossible. The whole thing.

Then the alternative stared me in the face—the sea.

It had to be possible, somehow; if the port engine burned up all its oil. When was the time to attempt this oil transfer? Now: or when we had evidence of the port engine failing?

I looked again at the outboard engines; away out from the fuselage, at the end of the strut: and I weighed up the chances, both ways. The chance of slipping, of being blown off the strut or the engine mounting, seemed infinitely greater than all my theories of running out of oil. After all, the engines were still roaring away at full throttle, and the only evidence of possible failure was the trail of blue smoke in the port exhaust. Perhaps I was putting it off, staying in the relative safety of the cabin: but I decided it wasn't worth it; unless the oil pressure began to fail.

The wind now had come more into the east, so, with some favorable component in its direction, we decided to alter course for Sydney. I gave Smithy the new course to steer and passed to John Stannage the necessary information for transmission.

For five hours Smithy had been flying the *Cross* in her disabled condition, concentrating for every moment of that time on keeping her in the air. He had lived and felt with his aircraft every effort of her struggle for survival. Knowing his feelings about the *Southern Cross* I rather diffidently suggested that I take over to give him a spell, and try to keep her in the air. He hesitated for a moment; then let me take her.

Immediately I laid my feet to the rudder bar and took the wheel in my hands, I realized the narrow margin by which the two remaining engines were holding her in flight. For a few moments I was lost in my endeavor to react to the needs of the aircraft; but gradually I began to pick up the sensitive signals, and finally to anticipate them and so to hold her in level flight a few hundred feet above the sea.

As I became more accustomed to the feel of the aircraft I was able to relax a little, and my eyes set off on the habitual round of the gauges on the instrument panel. The port oil pressure gauge, the danger point in my mind, was holding steady at 63 pounds to the square inch. Pressure on the gauge of the center motor was approximately the same. The needle of the starboard lay flat at zero on the gauge. The motors sounded healthy and I began almost to feel that the most critical situation was passing, as the engines burned down the weight of the fuel. We were able even to ease the throttles very slightly back from maximum power and still maintain height at three hundred feet. But my eyes continued regularly on the round of the gauges, and I still saw in my mind from the starboard seat the blue smoke trail from the exhaust of

the port engine. Apart from its numerical reading, I had noticed a small spot on the face of the port oil pressure gauge,
exactly where the needle was pointing. Each time I looked I
had mentally checked the holding of the pressure by the needle against this mark.

Now, when I looked again, my eyes were rooted to the
gauge and my whole body froze into a rigid warning. The
needle was flickering, and as it wavered about the mark on
the dial it was very gradually falling below that mark. The oil
pressure was definitely falling. No need now to be frozen
with doubt and anticipation. The port engine was obviously
close to the end of its lubricating oil; close to the end of its
life as an engine.

Feeling a dull and futile hostility, I attracted Smithy's attention and pointed to the gauge. A hardness came into his
expression as he took over his aircraft from me. He throttled
back the port motor, gave it several bursts, and then opened
to full power again. The pressure was down to slightly below
sixty pounds. We looked at each other across the cockpit
with an exchange of expression which obviously agreed,
"Well, it won't be long now."

I went below to the cabin, let Stannage know the situation;
and he immediately transmitted the signals, "Port motor only
last quarter of an hour. Please stand by for exact position."

I then worked up and handed him the estimated position,
which   he   transmitted,   "Latitude   34°8′S.,   longitude
154°30′E."

When I went up to the cockpit again the pressure was
down to 35 pounds, and Smithy was starting to take off his
heavy flying boots.

Suddenly all reasoning, fear, and emotion of any sort left
me, and were replaced by a clear feeling of elation; an obsession which listened to the promptings of nothing but itself:
"Get the oil from the starboard tank. Go out and get it."

I slipped below to the cabin, took off my shoes, belted up
my coat tightly, unlashed some light line from the mailbags,
and went back to the cockpit. Smithy was sitting there, flying
the *Southern Cross*, preparing himself to put her down in the
sea. I shouted across to him, "Going to have a stab at getting
some oil."

He shook his head and tried to stop me, but when he saw my determination he accepted it, and while we still had the port engine he tried to gain a little height.

It amuses me now to remember that I lashed the mailbag line round my waist and made fast the other end in the cockpit. It would have snapped with the slightest jerk, but it had a good moral effect, at the time. Then I stood on the starboard pilot's seat and put one leg over the side, feeling for the streamlined tube to the motor. The airstream grabbed my leg and for a moment a wave of futility swept over me. But it passed and again I was driven by the single purpose of oil for the port motor.

I finally got my right foot on the strut, held fast to the edge of the cockpit with both hands, and managed to get my other foot out, and hang on in the airstream. The blast from the center motor screamed round my ears and pushed with a numb, relentless force against my body. A wave of sudden panic surged within me and I felt the utter madness of attempting to move anywhere but back to the cockpit; if I could get back. I stood on the strut, with my shoulders braced against the rounded leading edge of the wing, with a screaming hurricane threatening to blow my eyes out if I looked ahead. Then the panic passed and I felt no sense of height nor any particular fear of the precariousness of my position: only again the obsession to reach the tank behind the motor.

I braced my shoulders against the wing and tried to wrap my toes around the strut; let go my right hand from the fuselage and edged my feet along till at the full extent of my left arm to the cockpit edge I found I could not reach the engine by reaching out with my right. I was horrified to discover that there was a short distance in the middle of the crossing to the engine where I would have no handhold and would have to move on out with only my feet on the strut and the back of my neck against the wing.

Momentarily, again there was a sense of defeat. It seemed almost certain that I would never make it, but just be blown off the aircraft and fall into the sea. Then I thought, well I'm going in the sea anyhow; so it's better to take a chance on reaching the engine. I braced my neck well against the wing, got a firm footing on the strut, and very carefully let go my handhold on the cockpit. There was an immediate impulse to make a desperate rush and grab at the engine mount; but I resisted that, and thoroughly steadied myself into the position without any handhold. Then I very carefully moved sideways towards the engine. Those few seconds seemed an eternity and the distance infinite, but I reached the engine mount, and clung to it with both hands. Then the worst feeling of panic of the whole operation swept over me—that of being isolated out there clinging to the engine with no way back but another horrifying foot-and-neck crossing of the strut.

But there was no time for panic. Smithy and John were making signs to me that the oil pressure was dangerously low and I knew something had to be done about it immediately. I hung on with one hand, and with the other tried to get the side cowl pin out so I could reach the oil tank. With maddening deliberation the pin resisted my attempts to undo it, but somehow my fingers dislodged it. The other pins came away quite easily and I wrenched out the side cowl and let it go in the airstream. Under the tank I located the brass drain plug.

I made signs to Stannage for a spanner, but he had anticipated this and by colossal luck had found a shifting spanner which we kept for dismantling the hand pump on the cabin fuel tank. I moved back as far as I could along the strut while still holding on with one hand; and with the other reached out to meet Stannage's hand with the spanner. The combined lengths of our arms saved me another passage without handhold. I slid back to the engine, got the spanner adjusted to the drain plug and eased it back till I could undo it with my fingers. Then I needed something for the oil.

Again John Staggage was ready. I saw he had some sort of metal container (which I afterwards found was that of a thermos flask he had for coffee). By the same process as we exchanged the spanner, I got the flask and quickly had it under the drain plug. To do this I had to hook one arm through the tubular engine mount, hold the flask in that hand and unscrew the drain plug with the other while sitting astride the strut. It was not particularly difficult really, but the airstream blew the oil away as soon as it came out of the plug hole. But I wangled the container up to the drain hole, got it full of oil, and put back the drain plug to a finger tight position. We could not afford to waste oil, with some hours ahead and the hungry port engine.

Now I had to get this container of oil back to Stannage. This we accomplished in the same way as passing the spanner and the container. After collecting and passing back to Stannage several containers of oil I had then to make the full return crossing to the cabin. I was fairly exhausted by that time so I cared less about the risk of the neck-and-foot crossing, and finally reached the cabin just about all in.

Stannage had been pouring the oil into a small leather suitcase which he kept for his radio gadgets and, again luckily, it did not leak. But the oil pressure was down to 15 pounds.

For a few minutes I simply could not move, or do anything but try to regain my breath. But that gauge got me on my feet again, and I climbed round Smithy in the port seat and tried to get my foot over the side for the passage out to the port engine. The howling blast of both slipstreams, center and port engine, hurled me back against the bulkhead and left me gasping and cursing in futile desperation.

Angry and frustrated by this setback, I looked out across the gap to the failing engine, still obsessed with the one idea of getting there. I forced my leg over the side and pushed with every ounce of my strength; yelled and cursed at the roaring flood of air; but was beaten back to the cockpit; stunned and defeated. Then I saw Smithy's hand go forward to the throttles and push them wide open again. He couldn't let her pick up speed to start an attempt to climb because it would have started the broken propeller windmilling. So he immediately hauled her back and willed and lifted her for height. He looked across at me as I still waited, gasping and hostile against the bulkhead; and I understood his intention.

At about seven hundred feet he shut down the port engine, leaving her flying at full throttle on one, and immediately starting to lose height. But this was my opportunity to reach the port engine, with its propeller now just whistling round without the blast of its powered slipstream. I went over the side and found I could force a passage against the blast from only the center motor, as I had done on the other side. I reached the engine just as Smithy shouted at me to hold on. I draped myself over the cowl against the V-struts and lay as flat as I could with my head behind the exhaust ring. The engine opened up again with a shattering roar, and looking down from my strange situation on the streamlined cowl section behind the engine I saw the gray surface of the Tasman only a few feet below me. The *Southern Cross,* flying only on one engine, had lost almost all the height as I was making the crossing to the engine. I lay on the cowl, not caring about anything but the temporary relief of not struggling against the airstream, and hung on with the breath being sucked out of my body, behind the roaring exhaust. I remember feeling something pressing against my ribs hurting terrifically, but it didn't seem to matter. There was only hanging on, and breathing, to consider.

Having gained a few hundred feet of height Smithy shut down the engine again. I had my back to the cockpit but it was obvious what he was doing to make it possible for me to transfer the oil. Relieved again of the worst airstream, I struggled up and attacked the cowl over the oil tank filler cap. It came away easily and I bent it back and was able to unscrew the cap.

Stannage was ready. He dipped a flask of oil from the case, I moved back along the strut and we both reached out till I took the flask from him and moved back to the engine. We lost a lot of the oil as it was sucked out of the flask by the airstream, but there was still more than half left as I reached the motor again and held the tin against my body. I climbed up into position over the oil tank, cupped my hand round the opening to avoid losing more oil, squeezed in the top of the flask and poured the oil into the tank. I looked back to the cockpit waiting for the reaction, but with just the ghastly thought now that it might not be shortage of oil in the tank, but a failing oil pump or a blockage in the system. But in a few seconds there was a great shouting and waving from the cockpit, and John Stannage held both his hands out with thumbs up.

Pressure! Oil pressure back on the gauge. It worked!

But Smithy signaled again to hold on. We were almost in the sea. I flung myself down on the cowl again and the motor came in with a booming roar. I could see the surface of the

ocean skimming by a few feet below: then I buried my head
from the torrent of air and waited for more height and a
chance to transfer the rest of the oil in the suitcase. As I lay
there jammed against the struts I felt a magnificent exhilaration and a reckless enjoyment of our success which made me
want to stand up and laugh and shout at the roaring mass of
air that tore at everything around me. In my mind I could
see the pointer on the gauge rise up and register the pressure
in the oil system. Then the pressure of the strut against my
ribs began to crush my body so that I began to feel that I
could not hold on any longer. The ocean seemed to be moving faster: then faster, and sinking farther away. A strange
ease and resignation came over me. Nothing seemed to matter. It was all some fantasy in a strange retreating background from which I was floating away.

Then a sharp stab of fear hit me and I realized I was letting go, and I felt again a choking numbness in my body, but
something telling me to hold on. Just to hold on; to fight the
unconsciousness into which I was slipping away.

Suddenly the roar of the engine ceased and I realized that
Smithy had throttled back and I had to get more oil. It
shocked me back to action and I lifted myself from the cowling and turned to move out and reach for the oil.

In a few minutes Stannage and I had transferred all the oil
in the case, about a gallon; but some had been sucked away
in the airstream. Then Smithy's shout came again and I had to
lie over the cowl again and hear the blast of the exhaust a
few inches from my ear. But I was past caring now, and
there was the exhilaration of knowing we could keep the
Cross in the air. When he had a few hundred feet of height
he shut down the engine again and I safely made the passage
on the strut back into my aircraft. My eyes went to the pressure gauge and I saw the needle at 63 pounds. Then I just lay
back on the big fuel tank in the cabin, and let go.

Stannage was again in touch by radio and informed Sydney that we were still in the air. That contact with the world
by radio seemed at first to give us some physical connection
with Australia, and therefore some basis of security: but one
quickly realized that the signals coming in through the wireless set were the faint sounds of a world with which we had

no connection, and only impressed upon us the vast solitude of our surroundings.

Fascinated by the oil pressure gauge, my eyes kept coming back to it for a reading, and I started to work out how long it would be before the oil transfer would have to be done again, and how many times it would have to be done in the distance we were still out from Sydney. Because I had lost so much in oil in the airstream, only a little over half a gallon actually reached the tank. The engine had burned eleven gallons in twelve hours: so about half an hour seemed like the limit of her endurance on the half gallon of oil.

I checked the speed, time, and distance made good, and estimated that the aircraft was still two hundred miles east of Sydney. We went over all the possible alternatives to this method of transferring the oil, and were forced back to the original conclusion that there was no other way to do it. We either kept on getting the oil or we lost the port motor and went in the sea.

In about half an hour I was horrified to see the oil gauge starting to flicker again. Till I actually saw it happening I had stayed in a kind of neutral state of mind, accepting the respite, and not really facing the fact that I would have to do it again. Now it stared me in the face. I made an effort to throw off all thought, and just act.

Again I reached the starboard engine, collected the oil, went out the other side, and finally completed the second transfer without incident. But I found that this time, to keep the aircraft out of the sea, Smithy had to tell John Stannage to dump the mail. It was a bitter experience for him, but it had to be done, to keep the *Cross* in the air, because now the port engine was occasionally misfiring and showing signs of packing up. Full throttle for more than a very few minutes brought ominous bangs from the exhaust.

And so we flew on, making the oil transfer about each half hour, throttling back the port engine to cool it off, and losing height: then bringing it up again and trying to gain a few feet on the altimeter.

About 120 miles from Sydney we sighted the smoke of a ship on the horizon, and later flew over her. (We later learned that she was a small New Zealand vessel, *Port Wai-*

kato.) Smithy spoke of putting the *Cross* down in the sea alongside this vessel, to give us a chance of being picked up: but I knew he was thinking this way so that I would not have to risk more oil transfers. Strangely enough, I had gained confidence in being able to go through this act without slipping or being blown off the airplane, and felt quite exhilarated at the possibility of our reaching Sydney and landing on Mascot airport in good shape. It was typical of Kingsford Smith that he was prepared to lose his aircraft rather than let me risk any more oil transfers; but I felt very sure of myself now, and preferred to go on getting the oil than deliberately to land in the sea with a "wheels down" aircraft. Had we ditched the *Southern Cross,* Kingsford Smith's chance of coming out of it from the pilot's seat would have been small. Stannage and I might possibly have made it, but it had no appeal to me and after a short discussion it was decided to proceed for Sydney.

Our real problem now was the port engine. Smithy had to cool it off by reducing power and each time he throttled it back we began to lose height, with the center motor still blasting away at full throttle. There was little point in worrying about the oil left in its tank. There was just no way of reaching it.

About three o'clock in the afternoon, while Smithy and I were both up in the pilot's cabin, we saw a low, purple streak on the western horizon. John Stannage came up, and our eyes never left this vision till we positively identified it as the coast of Australia. The sight of land impressed upon us the truly disabled condition of the *Cross;* but in nearly forty years of life it was one of the best sights I had ever seen. Now that we had actually seen the land it seemed infinitely far away; the aircraft seemed barely to be moving, and unlikely ever to reach it.

The intervals between the choking spasms of the port motor were closing upon us, and Smithy was forced to throttle it back every few minutes to prevent its complete collapse. Then it would cool off and gather strength for another burst, and respond again to the throttle, to keep us out of the sea. But the *Cross* had burned down most of her fuel now and was flying light, and gradually the land grew up out of the

sea till we were able to identify the higher land off the port
bow as the hills behind Bulli. The desolation of the sea began
to be more distant, though it still lay only a few feet below
us; and the world we had left in the night only fifteen hours
before began to creep back into my mind as a possible reality.

About thirty miles off the coast the engine was calling for
oil again and it was obvious that at 60 knots we could not
reach the land. Smithy was against my making this last passage for oil and again was prepared to put his aircraft in the
sea, since rescue, if we got out of the ditching, was almost
certain now. With a wry smile he accepted my suggestion
that we do the oil change and go right on in. We were quite
close to the land when the pressure gauge settled again on
sixty-three pounds. I watched the yellow sands of Cronulla
Beach come in and pass under the aircraft as Smithy coaxed
its last effort from the banging port engine.

With a perfect approach, he brought her in over the
threshold of the airport and feathered her onto the ground.
He turned the *Cross* from her last ocean flight, and brought
her to rest by the hangar.

The engine which had kept going at full throttle was the
one which had been strewn in pieces on the floor of the hangar and assembled mainly by John Stannage and Jack Percival.

Chapter 8

The Loss of Kingsford Smith

THE EFFECT of the circumstances which put the Altair out of the England–Australia race, and which in effect questioned the airworthiness of this aircraft in which we had since flown the Pacific, stayed in Kingsford Smith's mind as a challenge to the time of 2 days 22 hours and 54 minutes in which Scott and Black had made the flight from Mildenhall to Melbourne, and which then stood as the England–Australia record.

He accordingly shipped the Altair across the Atlantic to England to make an attempt upon this record. After the usual fuss and confusion with which earthly affairs invariably surround such a venture, Kingsford Smith was ready to go.

This time those disturbing influences were greater than usual, rising to an ever increasing crescendo of human reactions to the single, clear-cut purpose of flying the Altair out of England for Australia as quickly as possible. There was some opposition to Smithy making this flight, and I knew there was good reason for that opposition. He was a sick man: only partially recovered from influenza, but much more sick as the result of the persistent frustration of his plans to consolidate his pioneer work with the establishment of air services, and particularly the trans-Tasman service.

Just as the main pattern of shipping services had been set in the past with ruthless competition, that of the international air services was beginning to be laid down in similar circumstances. In this process the law of the jungle prevailed; and still does prevail today. Though my own clearly established purpose was and always has been exploratory flight, with no aims to operate air lines, Kingsford Smith challenged and continued to challenge some of the most powerful transport interests in the world, and he consistently lost. Had he and Ulm stayed together, I think they would have won, for each supplied an ingredient which was valuable for success in this jungle warfare.

But instead, Kingsford Smith, his gallant spirit never admitting defeat, gradually had his essential fiber whittled away, leaving effective only his spirit and his body with its unafraid smile. Kingsford Smith was killed in the jungle warfare of the fight for international air transport services. I saw this happening in the two years in which we were associated in our flying partnership.

When it was reported that Smithy had passed out when flying the Timor Sea in his Percival Gull on the last stage of his earlier solo record flight from England I was pretty sure it was not due to the carbon monoxide fumes which were said to have gassed him in the cockpit. He had recovered and gotten the machine under control with very little height to spare, flying on to Darwin and the solo record.

On the last of our experimental trans-Tasman flights in the *Southern Cross* he was overcome by this strange illness and in spite of his determination to go on he was forced to rest lying on the mailbags in the cabin for about an hour.

His associates in England tried to persuade him not to make this flight in the Altair. Mary Kingsford Smith, his wife, appealed to him from Australia: but it was no good. With him on this flight was Tommy Pethebridge, a faithful friend, a first-class engineer, and a pilot of sufficient experience to relieve Smithy in the air. Most of all he needed

Tommy to refuel and service the machine at the brief stops on the way, for that is where most fatigue originates if the pilot has to be involved in all this racket himself. Immediately the airplane comes to rest on the ground there is an invasion by all sorts of people who have some exclusive reason for engaging the pilot's attention and depriving him of the one thing he needs—rest. Smithy was wise to take Tommy on this flight. By allocating all ground contact to Tommy he would be able to get some rest, and charge his batteries, instead of having them finally run down by all the clamor and discord around him. The shock of descending into all this from the tranquillity and harmony of the air was well known to Smithy, as it is to me, and he wisely organized his flight to avoid it.

They left Lympne at 0627 on November 6, 1935, and the Altair sailed into the route to Australia in a convincing way. The Mediterranean, Baghdad, Karachi, were put behind him in good time and his last known landing was at Allahaband. Tired, he took off into the night for Singapore, two thousand nautical miles across India and the Bay of Bengal.

Somewhere off the west coast of Malaya he passed from this life. Those of us who were checking his progress in Australia were disappointed to hear that he was overdue at Singapore, and even when time passed beyond the endurance of the Altair, we were concerned at first only with the thought that he must be down somewhere and his flight time must then be outside the record.

But as time went on there was no news of his landing and we had to admit to ourselves the possibility of disaster. The Altair was only once sighted after leaving Allahabad: by Jimmy Melrose flying also for Singapore. Melrose, alone in a light aircraft, saw the lights of the Altair pass close over him out over the Bay of Bengal. It must have been the Altair because no other aircraft was in that position on that night.

Aircraft from Singapore and other points in Malaya went out to search for Kingsford Smith and Tommy, and all Malaya was alerted for news which might lead to their discovery in the jungle.

In Sydney we became more than restive. We had to go,

and search. Money was raised to charter a small twin-engined aircraft, and I was to take this machine and join in the search from Malaya. With Harry Purvis of Kingsford Smith Air Services, and John Stannage, I set off on the first day flight for Cloncurry in western Queensland. This was to some extent a test flight for the aircraft, before we pushed on the next morning to Darwin and through the Indies to Singapore. I was exhausted with the clamor and endless telephones and interviews before I reached the aircraft but I relied on the flight to restore my depleted resources. But it was a trying day. There were some problems with this aircraft and the air was rough for the long flight to Cloncurry. I sawed away at the somewhat unresponsive controls, keeping her on course, and we really worked our passage through the turbulent air over the hot lands of western Queensland.

At the hotel that night, far from the racket of Sydney, I planned to have everything organized and shaped up properly for the flight ahead. I would start with at least eight hours' sleep, to restore my own body to a going concern, make the day flight to Darwin, refuel, and go on through the night for Singapore.

I went to bed at eight o'clock, having given emphatic instructions that in no circumstances was I to be wakened or disturbed. I was not in very good form, having noticed myself being irritated lately by small things which normally I would have brushed off lightly. But I thought that the flying would soon smooth me out and restore my physical resources.

There was a personal problem behind this attempt to get an undisturbed sleep at Cloncurry. After the oil episode in the *Southern Cross* I had felt no bad effects from the experience; only a great relief, and relaxation, and a fine exhilaration from the success of our struggle over the Tasman. After the inevitable interviews at the airport, I had put together my navigation equipment, shed my oil-covered overalls, gone home and gone to sleep. (When these overalls were later sent to the cleaners they were returned beautifully finished and packed, with a little complimentary note "No Charge." It was a thought which I very greatly appreciated.) Because of the stresses of the whole affair, however, I knew it was important not to be disturbed in that sleep; to sink away into relaxation and let my whole body and mind gradually normalize themselves. So I asked my mother, with whom I was staying, not to allow anybody to wake me. A number of press people phoned for personal statements, and all, except one, respected her explanation that I must be allowed to sleep. But this man put up such a good story that he persuaded her it was very much in my interests to speak with him on the telephone, and that he knew I would be very upset if she did not wake me and let me know. She explained the importance of my not being disturbed, but he was so persistent and his story so convincing that she eventually believed him and I was wakened and brought to the telephone. I found of course that he merely wanted an exclusive story, and to get this he had persisted where all the others had decently accepted my need for undisturbed sleep. My reaction on being caught by this trick need not be explained.

But the breaking of the fine thread of relaxation that day was responsible for the problem which assailed me at Cloncurry and left me with a tough and persistent enemy on my back for more than a year.

Here at Cloncurry, with the thought of an undisturbed night ahead, I went to sleep almost immediately. The next thing I knew was somebody shaking me and saying something. I reacted immediately, thinking it was early morning and time to get up. But the voice said, in the darkness, "There's a phone call for you from Sydney."

Who could want me at this time of the morning, from Sydney?

"Who is it? What time is it?"

"Somebody from one of the newspapers; wants you to give him a statement about your flight."

I looked at my watch. The time was nine o'clock; I had been asleep for half an hour.

I don't know who it was who came with that message, but he didn't stay long. I was blind mad with rage. It was no good now. No escape from all this madness surrounding the one thing that mattered—that Smithy was lost and I had to get there.

All inclination to sleep had left me. I lay on the bed, trying everything. Got up and went outside. Even the stillness and the deep tranquillity of the Australian night had no effect.

Back to bed again, determined to empty my mind of everything. "Think of nothing. Think of nothing. There is only blackness. Nothing. Relax. Nothing. Let Go." I tried all the mental suggestions. Then I began to feel queer, to break out into a sweat, and then to feel that I was passing out. I got up off the bed, stood up, walked about the room to beat it. I felt sick, lay down again, and eventually dozed off fitfully, waking each time with a ghastly feeling of dread and fear of nothing in particular.

And so the night passed, the worst I can ever remember. In the morning I was drained of all life; but tightly strung and unable to relax.

I must get to the aircraft. I would be all right in the air.

Then I thought of the D.C.A. doctor at Cloncurry whom everybody knew and liked so much in this far place. Perhaps I could ask him to give me something to get me on the way. The air would do the rest. I called the doctor and he was able to see me immediately.

I told him my story; noticed he was looking at me and thinking beyond what I was saying. Then I asked him just to let me have something to start me off on the flight.

He looked at me in a kindly way and said, "You won't be flying today."

"But why? I must be in Darwin this evening."

"No, you won't be flying for some time."

A dreadful feeling came over me: a horrible realization that I was at the end of the road.

"I don't want to ground you officially, but you are not fit for flying. You must have a complete rest."

Rest. Rest, when my aircraft was fueled at the airport and ready to go. It seemed ridiculous; impossible. But he was quite insistent. I had to face it. I was medically unfit for flying and if necessary would be formally grounded. There was something utterly beastly about this. I felt unclean. Something abnormal and horrible, condemned to the earth.

I told him more, about the oil episode in the *Southern Cross*, how I had been completely unaffected afterwards, but had noticed myself being irritable lately. This only confirmed his opinion, that there was some delayed effect from this, as well as other things.

He eased me into acceptance of his opinion with great kindness and wisdom and soon had me thinking of going out to stay at a homestead at one of the stations near Cloncurry. This I felt I could do. Anything but return to Sydney where everything would be raked over and blown up into some hideous nightmare.

But one thing I knew I had to do: before I went to rest I had to fly this aircraft again. If I let it beat me I was sunk. I might never fly again.

The doctor surprised me by agreeing to this. So I went out with Harry and John, left them on the airport and took off. I had to have this situation out alone with the aircraft. I felt that at least part of my failure had been due to the fact that I had found no peace in the air coming from Sydney, and perhaps that was because I had not mastered this unresponsive aircraft.

The airplane seemed somehow to sense my thoughts, for she did not oppose me. I threw her around over the airport; did far more with her than I would normally have done; then slid her in to land, and stopped at the hangar. That was all right. I would fly again. I could look this airplane in the face with a friendly smile.

I went back to the hotel, got my things, and went out to the property in the car which had been sent for me. There I knew for a fortnight the most wonderful hospitality. In spite of the disgrace which I still felt within me, I was happy. Nobody there ever referred to my reason for being with them.

But how slender is the thread by which our fortunes hang. Harry Purvis took the aircraft back to Sydney. About twenty-five minutes out of Cloncurry, when he was flying very low in the smooth air of the early morning, he found he had no lateral control and the aircraft would not respond or correct laterally with the rudder alone. Straight ahead was a long smooth clay-pan. Just in time to avoid an uncontrolled bank with fatal results he drew off the throttles and put her down to a perfect landing.

Having taken a deep breath and looked around in appreciation of his position he examined the controls and found that a cable had slipped off a pulley and the aileron control cables were loose and quite ineffective. He put the cable back, tightened the turnbuckle, started the engines and went on his way, to Sydney.

Had we gone on for Darwin this would have happened at about seven thousand feet after the climb out of Cloncurry, and it is very doubtful whether it would have been possible to retain control of the aircraft. So, destiny goes along with us, and I was grounded only for a fortnight, when it might well have been permanently.

Some time later a searcher from Malaya found an aircraft's wheel on the beach of an island in the Mergui Archipelago. This was positively identified as the wheel of a Lockheed Altair. Only three of these aircraft were built. The other two were out of service: so there can be no doubt that it was somewhere in that region that Smithy and Tommy went in.

The pilot of a small aircraft, who had been active and persistent in the search, reported some time later that he had seen a gap in the treetops of a very small and precipitous island where something had cut through the trees and from what he had seen he was convinced that this was where the Altair had crashed. It was close to this spot that the wheel was found.

Whatever happened was sudden and unexpected. I think it most likely that when straining to see something in the darkness as he made his low approach to the land, tired and perhaps completely finished after all that had happened before and in London and the long continuous flight without sleep, Smithy's body had just given out as it had done before, and the Altair had nosed in before Tommy could do anything about it.

That he should never have undertaken this flight is a popular and general opinion. The circumstances that led him to the flight forced him to do it. As it was, there was no other way for Smithy. To have dragged out his life in some physi-

cally secure but drab situation would have been death for
him anyhow. He was completely right in setting out upon this
flight. It was necessary, for the freedom of spirit upon which
he lived.

Chapter 9
The Way Ahead

BACK IN SYDNEY I faced a complete anticlimax. I had failed in the search for Kingsford Smith, and though I felt perfectly fit and wanted to fly, I could not sleep if I were going to fly the next day. I had at this time one of the most responsive and beautiful machines I have ever flown, a slim-bodied Percival Gull with a Gipsy Six engine. The Gull was a dream airplane, in general character very like the Spitfire of later years, but of course with nothing like the power performance. I looked to my blue Gull to reinstate me in the air, but every time I was going to fly her I would lie awake at night, with no fears but the fear of not being able to sleep.

It was ridiculous. I wanted to fly, yet every time a flight came up some subconscious control took over and kept me awake and staring into the night. It began to be really serious because I was not gaining ground. I fought it out for weeks. Then I realized that there was some influence there which would have to be located and destroyed.

I told my story to the doctor and, like the doctor at Cloncurry, he was watching beyond the words I gave him. At the end I could feel it coming, from his kindly sympathetic manner.

"How long have you been flying?"

"For nearly twenty years. I started in 1916."

There was a pause, and a smile, in confidence. "Haven't you had enough? Don't you think it's time you took a more important job on the ground?"

This was heading the wrong way altogether. "No, I don't see it that way."

"Well, suppose you ease up. Can you get some easy flying, where you won't be under the same strain?"

No good. I could see it was hopeless. I wasn't getting anywhere. Not where I wanted to go, anyhow. The rest of the interview was a polite tailing off. I left and went out into the street. Depressed and frustrated I went on back to my club, and sat down in a quiet room to assemble my thoughts and make some sort of positive decision and plan.

In my mind I looked at the Gull, and from her to the country of Australia, the coast with the beaches, the mountains and the plains, out to the red mountains of the Center. There must be work for a fast aircraft of this sort in Australia. What I needed was flying. So much, and of a sort that it would kill this anti-sleeping bogey. I needed to fly so much that my subconscious mind would become so reconditioned to the air that it would accept it as its natural element. Then I would just have to sleep from necessity and familiarity. I needed to fly so much that the present would wipe out the past as an influence in my mind, or at any rate so override it that the present would be the dominant influence.

Kingsford Smith and Charles Ulm were gone. The other flights for which I had trained as a navigator were before me now: but I could not undertake anything of this kind till I had completely reinstated myself.

I confided my thoughts and plans to Jack Percival, aviation correspondent of the *Sydney Morning Herald*, who had been official correspondent in the *Southern Cross* on my first flight as navigator from Gerringong Beach to New Plymouth, and was therefore a good subject for the adventure of flying with a pilot who couldn't sleep. We conceived the thought that there were from time to time news events the details of which the *Herald* would surely like to have for publication far ahead of any other newspaper. With the Gull, we could give the *Herald* a scoop on some of these, by flying-in the news and the pictures faster than the airlines which were normally used for transportation. I foresaw clearly that there would be some tough experiences in these operations with the Gull, because I was bound to run into situations where I could not decently escape from the sort of flying which was against my principles. I also saw that by comparison with the obvious demands which would be made upon the Gull—and if I could survive them emotionally—ordinary flying would afterwards be a fairly straightforward sort of affair.

We put this idea to the *Herald* and it was well, if cautiously, received. I could see that much would depend upon the results achieved on our first operation.

The first assignment was to deliver pictures of Australia's most famous horse race, the Melbourne Cup, which was run after the last regular air service of the day had left Melbourne for Sydney.* By flying at night we could deliver the pictures to both the *Herald* in Sydney and the *Courier Mail* at Brisbane for publication in the next morning's newspaper, thus scooping the next evening papers which otherwise would be first with pictures of the Cup.

The morning of the race Jack Percival and I flew down to Melbourne in the Gull. It was his job to collect the pictures from the photographers on the racecourse and to deliver them into the airplane at Essendon airport: my job and the Gull's to fly them first to Sydney, and then on to Brisbane.

The race would be over shortly before four o'clock. Jack Percival was due back at Essendon from the racecourse at 4:15, by a picture delivery and transportation system in the best dramatic movie style.

By 4:10 I had the motor running and warmed up ready for take-off. Four minutes later a car screeched into the airport and drew up on the tarmac alongside the Gull. At 4:15 Jack was seated in the aircraft, holding the spoils of his raid on the racecourse. A minute later we were in the air for Syd-

---

* At this time there were no night air services and no facilities for them.

ney, the Gull climbing and reaching for distance with an eager, rushing sound of air and roaring motor.

I put her on the direct course for Sydney while Jack sorted out the various packages of pictures he had captured from the photographers.

At ten thousand I leveled her off and let her ride the westerly out over Mount Buffalo, west of Kosciusko, and by the Bogong Peaks; and as the sun was sinking low into the dust haze on the western horizon Canberra came out from under the port wing.

We saw little else till the lights of Sydney came in close below and a few minutes after seven o'clock I rolled the wheels onto Sydney airport. The *Herald* man collected his pictures and the Gull won her first bet with an easy swing over the southern sky.

In a few minutes the fuel truck was alongside and the line delivering gasoline to the tank. We had some coffee and sandwiches and were ready as the tank caps were secured. The Gull hurried off into the night, with the pictures for the Brisbane *Courier Mail*.

Under the dust haze Sydney lay in a blur of dull red light, and the shore of the harbor was barely visible as it passed directly below us. I laid a course for Nobby's Light at Newcastle, allowing for drift on the westerly, and let her go on, climbing. The air was hot and turbulent but the Gull rushed along through the night, enjoying it. We flew with her, enjoying it too, with the *Herald* job chalked up behind us and the night air inviting us on ahead.

To be sure of arriving over Brisbane with plenty of range in reserve I had arranged to land at Coff's Harbor for fuel. In still air the range of the Gull would put her over Brisbane from Sydney with a few gallons left in each tank. This night the weather system over eastern Australia showed a probability of fog on the north coast and at Brisbane airport late in the night: so I wanted reserve fuel to nose around in the night for somewhere to land if Brisbane should be under fog with zero visibility. Including the Coff's Harbor landing we should reach Brisbane at 11:30 P.M., with a good time margin for publication of the *Courier Mail*'s pictures.

For much of the flight up the coast we could see nothing

but an occasional star, only faintly visible through the dust haze which had come over on the high westerly from the dry lands of Central Australia. But north of Port Macquarie the haze began to thin out and we could see the difference between land and sea. In clearing air we passed the light of Smoky Cape and a quarter of an hour later came in over the coast for Coff's Harbor airport. It was just possible to distinguish the flares, flickering dimly through a ground mist. This was a trap, for though the line of flares can be seen from vertically above, as the aircraft comes in low on the final approach to land, horizontal visibility may be zero as she actually enters the ground mist with flaps down and everything set up for landing.

I laid the Gull on an easy turn, circling the airport at a safe height from the invisible hills below in the blackness, and weighed up the factors for and against a landing. In these days I had of course no radio for communication, so would not know the conditions at Brisbane until actually arriving over the airport. If then, with only a few minutes' fuel remaining in the tanks, we should find only a ghost-white floor of fog below us, we would be in trouble.

The dimly flickering flares below seemed more acceptable. I decided to land; refuel and go on.

I eased down the power and let the Gull sink into the night, turning in from the westward and lining up for the flares. She came on in, almost silent now, but alert and watching, in the still air of the night. I held her on the line of flares, just visible now and flattening as she came in low to the invisible ground. Then suddenly there were no flares—nothing, but the Gull in my hands and the luminous instruments staring at me from the panel. In the ground mist. Power on and climb away, or go on with the landing? On, with the inevitable landing. Seeming almost on her nose the first flare came in, visible again, and others stretched away into the mist. It had to be now. I drew off the last of the power and she touched. I heard the patter of the grass and rushes brushing the spats over her wheels and all my body seemed to flow with the Gull in a smooth and wonderful sense of relief that she was down and rolling smoothly by the flares.

Crash! I felt the Gull bound into the air. Instinctively I rammed the throttle open to pick her out. The motor, stabbed in this unusual and merciless manner, responded, hauled her along in the air to prevent her dropping flat to the ground, and let her slowly down again. She swung round in the darkness and came to rest with the wingtip almost in the ground and the tail down at a horrible angle.

The *Courier Mail* pictures! My Gull, lying smashed and hurt on the ground! A dreadful feeling of despair swept away the relief of the landing.

I swung open the door and leapt out into the mist-wet grass. People had already gathered and were standing silently around, just looking, in the weird light of a wavering flare close to the machine.

Car lights swung round and came glaring towards us. I was angry and hostile now. Something on the flarepath. What had she hit, right on the landing run? But, with perfect judgment of the tension, nobody asked me what happened. Jack brought the torch and we examined the airplane. One tire was flat on the ground, the spat torn and twisted back under the wheel: an apparently hopeless mess. The end of the fuse-lage was on the ground, the tail wheel gone, the fairing buckled and jammed against the rudder, locking it. I looked at the time. Nine-thirty. Still time to reach Brisbane if we could do something about this mess.

There was no time for regrets and recriminations now. We had to get going somehow. I asked the oil company's agent to get the fuel aboard, went round to the undercarriage leg and with plenty of willing help set about trying to free the fairing from the flat tire. We managed to get the fairing off and ex- amined the undercarriage leg. Apparently it was undamaged. If we could repair the tire the airplane would be standing square on her undercarriage again, minus the spat but flyable so far as the landing gear was concerned.

Round at the tail it was different. The tail wheel was gone altogether, broken off at the post. There was nothing we could do about that. I looked at the rear end of the fuselage, resting on the wet grass. It was strong and not too vulnerable for slithering over the grass. We could do one take-off and landing on the fuselage, without a tail wheel: if the tire and tube could be repaired.

We managed to get the wheel off and drove into Coff's Harbor town to find somebody who could make this repair. I called the *Courier Mail*, told them the news, and said that we would try still to get the pictures through in time. But the tire expert, wakened from a peaceful sleep and confronted with our problem, looked seriously at the remains of the tube and I had the ghastly impression that in his mind he was already viewing the dead body of this indispensable equipment. But I knew enough of experts not to hurry him. After allowing a decent time to elapse while he examined the many cuts and punctures in the rubber tube which would decide success or failure to our venture for the *Courier Mail*, I faced up to the grim question and asked him, "Can it be mended?"

This man had the quiet, unhurried confidence of one who knows his job. I felt that whatever he said would be final and inevitable; a statement of fact, grim or hopeful, which would have to be accepted. With maddening deliberation he counted the holes, made a final appreciation of the situation, and answered my question: "Yes. It can be mended."

My hopes soared.

"You mean it can be mended here—now: that we can get away tonight?"

"Yes."

"How long will it take?"

"About ten minutes for each hole."

Impatience flashed through me.

"Can't you mend them all together?"

"No. We can only vulcanize one at a time."

This, I could see, was final. I would simply have to wait and watch with appropriate respect while the intricate rites were performed with patches and steaming presses and very obvious skill.

The repair was finished a few minutes before two o'clock.

Out at the airdrome the ground mist had now thickened into fog. The airplane, a dim shape in the night, lay inert and lifeless. An eerie fire burned near the aircraft, which was surrounded by crouching figures warming their hands. The flares had gone out, and beyond the Gull and the fire there was darkness; but I knew that a take-off into the northeast with the gyro set to the compass would put us over the sea with no obstructions to clear.

To avoid unnecessary damage to the fuselage, I asked several of the spectators if they would help us by lifting the tail and running with it as long as they could while the machine accelerated. If they could do this, even for a few yards, there was a chance that I would have enough air control from the elevator to keep the fuselage off the ground. This idea was received with much amusement, but they agreed to try it and we were ready to go.

I faced the machine in the direction for take-off, took everything out of the rear locker, got as much weight forward as possible, and Jack leaned forward with trusting confidence in this frightening take-off. I warmed up the motor, and the fog, though enveloping everything ahead, thinned a little above us, drifting in thin wraiths across the moon.

I checked the gyro heading with the compass, signaled to the tail-lifters, and pressed the throttle lever forward. I don't know how long they managed to stay on the tail, but there seemed to be only a moment when it sagged and brushed the grass. Then the Gull hurried away into the night, the motor spinning and calling her to flight. I wound down the tail trim and in a few moments she was airborne and away up through the mist, climbing, for the moon.

A few lights glowed weakly on the coast below the port wing, and the struggle on the ground became a distant thing. The motor sang again in tune with the night, lifting us up for the clear world above. It seemed to turn to us and laugh with a gay abandon, like children running with wild things through enchanted woods—escaping, out and away, with distant voices calling, moving swiftly, lighter—flying, with the stars.

I looked at the time. Only 2:30. We should be over Brisbane airport, Archerfield, at 3:45, and here I knew that Andy Laughlin would have out flares. Since the regular airlines had not yet introduced night flying services, the flares at Coff's Harbor were a private arrangement and those at Archerfield would be put out for us by the Department of Civil Aviation officer in charge of the airport.

Far below us now, down through the thinning haze, a faint silver light showed the surface of the sea. We passed inland near the Clarence River, then north by a few scattered lights of towns. The darker shadow of Mount Warning crept by the wing and the Macpherson Range came in below us.

Ahead was the low land by Brisbane, covered completely by a sheet of cloud, a flat layer of white shrouded by night to silvery gray under the moon. The weather system had fulfilled its promise. This was the earth cloud, steamed into life by the opposing temperatures of land and air, lying inert upon the world below us and completely obscuring it from our sight. I checked my reckoning. In fifteen minutes we should be over the airport. I peered ahead, searching for a gap. Then away ahead over our starboard side the silver sheet seemed suddenly filled with flashing lights. It was Brisbane, below the thinning screen of cloud.

In a few minutes we were over the edge with clear air below, black earth studded with scattered lights, and out ahead over the port wing was a straight line of wavering lights; the kerosene flares on the airport. I opened the cabin window and let cool air rush in with the smothering roar of the motor. As the lights came abeam I turned the Gull up on her wing and she sailed around the circuit.

The green Very light signal burst out against the ground, came slowly up like a comet sweeping through darkness below, to curve, and fall burning back to earth. Clear to land. With a feeling of high elation I brought the Gull round, lined her up on the flares, and sank her in. As the wheels touched the grass I let her run, holding the tail off the ground with the elevator, till the slowing airstream could hold it no longer. I held her with the brakes and let the tail sink onto the grass.

The *Courier Mail* man was there. We handed over the pictures.

As daylight came over we drove in to see the final processing at the *Courier Mail*. At seven o'clock the papers were on the streets, with the pictures of Wotan winning the Melbourne Cup. Between us all we had scooped the evening papers and won.

After repairs at Brisbane, we landed at Coff's Harbor on the run back to Sydney. The tail wheel had been retrieved from the long grass on the airport and we discovered the cause of the damage to the Gull. A log had dropped off a timber truck driving across the airfield and had lain, unseen, in the grass by the flare path.

The next charter from the *Sydney Morning Herald* was a series of flights to deliver pictures of the cricket test matches. In public interest these almost equalled the Melbourne Cup. Play usually finished at the end of the day, well after the last regular air service had departed for Sydney. Matches were played at four of the capital cities of Australia: Sydney, Brisbane, Melbourne, and Adelaide; so for each of the matches at the last three cities the *Herald* needed fast transport of pictorial news of the play, particularly the last of the play.

We ran a successful series of flights with the test match pictures and the last of these, from Brisbane, stayed particularly in my mind.

The Gull took off from Archerfield soon after the close of play, on a direct course for Coff's Harbor. From the Macpherson Range on this sublime and beautiful evening the western pinnacles stood against the sunset proclaiming some deep and ancient mystery. Seen from the ground, they stand mysterious as the stars against the night sky; but flying close by these great sandstone masses, their rocky faces bathed in blue and orange lights, I had the feeling of sharing with them the freedom they have achieved, ageless with time. Their spirit inhabited the air rather than the land, giving this impression of wild freedom; but there was something awe-inspiring about these great towers of rock. They gazed serenely at the aircraft, making it seem a short-lived thing, but their manner was friendly as they watched us pass, telling of eternity. As we flew away, the blue shroud of mist crept over these sentinels of time, leaving them distant again; again mysterious as the stars.

In the last clear light of an early high-pressure sky, we ran quickly down the two hundred miles to Coff's Harbor, and landed there for fuel.

The red and white tanker was alongside as I switched off the engine and the propeller fell over the last few compressions. Filler caps back, funnel and line to the port tank, and the fuel began to flow. Long shadows of the Gull and the fuel truck stretched across the airfield as the last of the light sank into the ranges behind the coast. The gauge swung in the glass housing on the port wing and I tapped it to be sure of an accurate reading—full. The line went over to the starboard wing and fuel ran into the tank; cool, clear liquid soon to be changed to fire in the flaming furnace of the cylinder heads.

Both tanks full. Caps run down, and tight.

She fired on the first swing. I passed a hand to my friends on the fuel truck, slid into my seat, and taxied out for the take-off. In a few moments we were airborne and I laughed with the aircraft as the Gull flowed along in still air, low over the darkening trees, the sandhills, and down the swing of the beach as the surf rolled close under her wing.

I saw the flash of the light on Smoky Cape; and ahead a layer of scattered cloud lay over the land with tops at about four thousand feet. I let the Gull reach up to rise above this layer and give me a sight of the distant sky far ahead.

When she was well clear of the cloud I turned down the instrument lighting and peered into the purple haze away in the southwest. I fancied there were the dark outlines of weather but there was not enough light to be sure so I dismissed it from my mind and settled back in the seat.

The motor ran with that perfect, spinning rhythm that with the stillness of the air emptied the world of everything. The land below, the black holes in the cloud, were no longer real; only a darkness under the cloud: nothing. Only now where the Gull flew was there life of any kind; life in sound and light when the brightest stars shone above a sky still faintly colored in the west, and the strange but familiar music of space came in on the drumming of the motor.

The shadow of night moved over and there were stars, gray cloud tops, and darkness. Something flew with us, near, in the cabin of the Gull. Now there was no distance, time, or space. Just sound and something flying with us, through eternity.

A light flashed away ahead on the port side, low in the blackness below the last fringe of cloud. Crowdy Head, beyond Camden Haven. I saw the faint lightness of the inlet under the greater darkness of the twin mountains at the Haven. In a few minutes the light came round by the port wing and I had an impulse to pull her over and go screaming down to shoot it up.

Away in the distance another light came over the sea. I checked the flash. Cape Hawke. Then suddenly the light was blurred. I looked up and the stars were gone. Rain rushed against the screen and my eyes instinctively went to the flight instruments. I transferred my night vision to the panel and settled to ride the Gull on the more intimate information from these luminous friends in the cabin.

But in a few minutes we ran out of the rain and I searched the night ahead for more distant lights to check the track of the aircraft. Stephens Point, Nobby's at Newcastle, and Norah Head. I altered course to pass west of Cape Hawke and with relief saw the flash of Sugarloaf Light, but it was Norah Head that I wanted. When we were abeam of that light there were only fifty miles of uncertain air between it and the airport at Sydney. As we passed the Myall Lakes the glare of Newcastle rose over the hills to the southwest, and Nobby's light sent out its group flashes. Far to the southward, blessed relief, was the sweep and flash of Norah Head, seventy miles away. Other nights I had seen its beam with more casual interest. Tonight, with rain and cloud about, it gave me safe passage to within fifty miles of Sydney.

Below, bushfires spread over the land, like torches in the darkness of space. I sat comfortably in the seat now, with Newcastle clear and Norah Head in sight.

But this peace of mind was short-lived. In a few minutes there was only blank darkness ahead. Whatever was coming in from the south was moving fast and wiping out the few pinpricks of light south of Newcastle. I ran her down to a thousand feet for position to take her under it if there was any visibility, and waited, with the uneasy feeling of going into some unseen action. Newcastle showed up again beyond the starboard wing. I altered course to 198 degrees on the compass and turned her nose to darkness in the south.

Sudden savage waves of air buffeted the wing in violent turbulence as she hit the edge of the front. Thin wraiths of cloud chased across the stars, closed over the cabin top, thickened, and shut us in. For a moment, lights at the entrance to Lake Macquarie showed faintly through a shower. Then heavy rain hit the aircraft, shrieking on the screen, and blotting out all vision ahead. Back on instrument flight, I tried to interpret the needs of the Gull for trim and course, as she slathered her way through the turbulent ocean of air and cloud and streaming rain. Drips of water came in and fell upon my leg, strangely connecting us with reality in the world outside, but the faces of the flight instruments came through my mind directly to action through my hands and feet upon the controls. I eased down the power as a violent updraft sought to suck the Gull up into the belly of the cloud, and pressed her down to oppose the rate of climb. I was all out to keep control of the aircraft, and hold her on the course: so for a moment I did not interpret the meaning of the light that swept through the blackness outside. Then suddenly, a miracle: clear air and the flash of Norah Head close ahead.

I laughed and sang again at my reactions of a moment ago, dived her down, and roared along the land. The lighthouse went by and she flew into the last stretch for Sydney. Tuggerah Lake was clear, but there was no sign of Barrenjoey light nor the glare of Sydney. I went for some height again, leveling off to clear the high coastal hills with a few hundred feet to spare.

There was a brief sight of Barrenjoey and Broken Bay but where the lights of Sydney should have been there was only a dull coppery glare shutting out everything. Then, the sound of rain again. I was tempted to let down lower for sight of the ground as Sydney must be coming in, but obeyed this impulse only to the limit of absolute safety on the altimeter. She swung and stabbed at the broken air and the rain roared above the sound of the engine. No good. I must hold the height, and if nothing was visible before the time allowance for Sydney airport ran out, turn east and somehow make contact over the sea. She seemed to be going at terrific speed, just clearing the hidden hills and housetops in my imagination. Things were just missing the wingtips it seemed, all hidden in the darkness as the rain shrieked and tried to tell me we were diving for the earth. I had to deliberately believe the flight instruments, to resist the calls of chaos outside the air-craft. Four minutes to go, and then I must turn east for the sea. But the scene changed with dramatic suddenness. In al-most a blinding flash the lights of Sydney were spread out below, clear as crystal and flashing like all the stars in the heavens thrown down to cover the earth below the Gull.

She spread the shadow of her wing across the lights and ran them under the nose, drove on through the open night and gave me a sense of triumph and exhilaration. I realized that sweat had been pouring from my head and I took a handkerchief and freshened up for the landing.

The Gull streaked across the brilliant city, over headlights of cars stabbing the wet streets with shafts of light, electric signs playing on the hoardings, dark patches of the parks and rows of lesser lights that marked the streets. All these things, which a moment ago would only have been a fantastic wan-dering of the imagination, were real and I felt an exhilarating triumph which made the motor spin and the wing ride on the air, calling to the night.

Another squall was coming in south of Botany Bay. I flew directly over the airport and turned her up waiting for the green signal to land. A red light rocketed into the air, curved, and fell burning to the ground. I watched impatiently as the squall approached, but with a small margin to spare the floodlight spread over the grass of the airport and the green signal broke out in the air. I shut down the motor immedi-ately and brought her in. Flaps down and a little power as she came in on the final approach. The shadow of the Chance light flickered under the port wing and she floated across the grass, weird and unnaturally brilliant in the artifi-cial light. Throttle off, and the ground took her as she lost speed for flight. The Gipsy beat with a pleasant, bubbling note of ease as she taxied in, round the tarmac, to her home in the hangar.

Only then I was really conscious of having pictures for the *Herald*. I handed the package to the waiting courier and his car swung out of the airport, and away.

And so the news events came up—and the Gull went out. Darwin Cyclone. Coronation pictures. Rabaul Earthquake.

Hurrying north or west over the Blue Mountains and far across Australia. Over the Timor Sea and through the Indies to Batavia: back with pictures of the coronation of King George VI and Queen Elizabeth carried from Europe by the fast K.L.M. Douglas to the Indies. A stop for fuel in the heat of Koepang: then back over the Timor, across the Kimberleys and in to land on the salt-flats outside Wyndham. Refuel; and on into the night for Alice Springs in the dawn. The night passage across Central Australia. Clmbing high over the southeast trade for a fair wind on the flow back from the west. Absolute darkness below and the stars above in a moonless night; steering a course on the compass and estimating Alice Springs from belief in this northwesterly wind at altitude. Excited and tense with expectation of the landfall at dawn. Venus rising in the night, so brilliant that a faint ethereal light came over the land. And with the dawn the outline of Mount Barkly, the jagged backbone of the Reynolds Range, and over the yellow glare of the saltpans bluemisted Mount Leibig in the Macdonnell Range. Then . . . the white roofs of Alice Springs. Fuel again for the tanks, and away for Broken Hill.

Other flights, emergency calls to the country carrying specialist doctors from Sydney; gold-seekers, out to the ranges of Central Australia; any flight within the capacity of the Gull. It was good, flexible flying: and sleep came in many strange places and circumstances. Full nights in the comfort of hotels; snatched hours or minutes anywhere. Sleep in strange places. But sleep. The Gull had done it. After six months with her I could sleep anywhere, with the prospect of any flight before me. I had taken the cure: intensive flying, till it became again my natural state.

Chapter 10

Cocos, the Elusive Atoll, 1939

IN JUNE of 1939 we made the first crossing of the Indian Ocean, and a survey of island bases which soon afterwards were used by the Royal Navy and the R.A.F. for operations against enemy submarines and surface raiders.

The purpose I had originally conceived for this flight was to give a lead to direct air communication between Australia and Africa, and to explore island bases for a trans–Indian Ocean air service. The idea of an air service between Australia and Africa was, however, so far beyond the political and public view of the times that to obtain support for this flight I knew I would have to "sell" it on a much more visible objective.

In my submissions to the Australian Government I therefore stressed the need for a reserve air route across the Indian Ocean which could be used to maintain air communication with the United Kingdom in the event of the Singapore route being cut by war, and proposed that we should make a survey of island bases over the region of that route.

Through Mr. R. G. Casey (then Treasurer and now Lord Casey) the interest of the Australian Government was aroused as early as 1938. But to find a suitable aircraft, or an aircraft which could fly across the Indian Ocean at all, was a problem which had no solution in Australia.

In England, however, the great days of the flying boat were just beginning, for both civil transport and overseas service operations; so I went to London in search of an aircraft with enough range to fly the longest stage of the Indian Ocean crossing, between Cocos Island and Diego García in the Chagos Islands.

The result was disappointing. The performance requirements of British flying boats was for short- and medium-ranged types, and though it might have been possible to tank up one of these aircraft to fly the distance, with little fuel reserve, I wanted something that would stay afloat in the air for really long endurance to cover the possibility of having to search for any one of the small islands on the route.

The next possibility was the United States, where I knew that Sikorsky, Martin, Boeing, and Consolidated of San Diego were well advanced in flying-boat design. I went on across the Atlantic, and a survey of American types revealed that the P.B.Y. long-range patrol flying boat then under construction by Consolidated measured up in every way to the type I needed. The P.B.Y. had a range of more than 2500 nautical miles, it had very reliable engines, practical auxiliary equipment; and most important, there was a chance that one might be available somewhere down in the region of New Guinea.

One of the first P.B.Y.'s off the production line was acquired by Richard Archbold as transport for an expedition to New Guinea and other places in search of fauna for the American Museum of Natural History.

With Archbold as my objective I went on home down the Pacific and finally located him in a remote part of New Guinea, by cable from Sydney. Soon afterwards his aircraft, the *Guba*, flew down to Australia and I was able successfully to negotiate its charter for the Indian Ocean flight.

The timing for this charter forced me to accept a rather horrifying financial risk. Archbold was making his plans to return to the United States, *Guba* was leaving again for New Guinea, and to be sure that I would not lose this aircraft I had to come to terms immediately with Archbold's representative. I had no promise of support from the Australian Government, merely an awakened interest through Mr. Casey; so, with a shaking hand, I took on the charter myself, well knowing that I hadn't the sort of money personally to fit such a venture at all reasonably into my private affairs. I had already spent rather a lot of money running down a suitable airplane around the world, but I had to accept the risk of this charter to strike at the right, and possibly the only, moment for success.

It proved to have been the right move.

With a world war threatening in Europe, and likely to spread in unpredictable dimensions throughout the world, the need for a reserve route to the United Kingdom and operational bases in the Indian Ocean was appreciated by the Australian Government. Charter of the *Guba* fitted very well into the needs of the times and it was not long before the Australian and British governments joined in taking over my commitment and appointing me to undertake the flight and survey.

To increase public interest in an air service between Australia and Africa I planned to carry an official air mail and to have a special stamp struck for the flight. This was approved by the Postmaster General, endorsed by the government, and the commemorative stamp was designed and a few were produced. Just as the main issue was to be available to the public, however, I received word that there was some problem about the whole question of the mail, allegedly because it was considered to be against postal procedure; that in fact the Australian Post Office would not now agree to the mail.

I immediately communicated with Mr. Casey and it was not long before I heard from him that all was well. The mail would be reinstated. This sort of thing was quite routine, not necessarily regarding special mails, but in the whole matter of pioneer flights, part of the impact of which was to alert commercial air transport people who imagined their personal interests might in some way be affected by the flight. I knew that somewhere along the line somebody was anxious about the implications of our plan to make the first air crossing of the Indian Ocean and to stimulate public interest through the mail.

Having, as I thought, disposed of this temporary setback to our mail, I dismissed it from my mind and got on with the job of preparation for the flight. But a week later Mr. Casey called me from Canberra and told me that we were in real trouble about the mail. The Kenya Post Office had now refused to accept it at Mombasa.

It was obvious now that somebody was really bothered about this mail. The Kenya decision was quite firm and representations to London were unable to change it. I was not concerned, however, with the interests of any airline. I wanted to create the maximum possible public interest in the idea of an air service between Australia and Africa and I believed the mail was an important factor in this interest. So I asked for authority to divert from our Kenya objective on the East African Coast and to make arrangements with Portuguese East Africa to go in with the Indian Ocean mail to Moçambique. I knew that this outrageous suggestion would have considerable repercussions, ones which might force Kenya to accept the mail.

But the Australian Government could not go along with this plan. Australia had not at this time achieved the inter-Commonwealth independence of spirit which has now released the country to such prosperous and active development, and there were real difficulties for the government in my suggestion; but I could see nothing but the objective and found it hard to accept the cancellation of the mail. When it was definite and irrevocable however, I dismissed the whole affair, and concentrated on the operational details of the flight. I then made arrangements to carry an unofficial mail and subsequently had the covers stamped in a Kenya post office. We distributed some to interested people, but the effect in public interest was quite small compared with that which would have been achieved by an official mail.

With me as second government representative in *Guba* was Jack Percival, who had traveled with me much of the way to the finding of an aircraft, and whose experience on some of the trans-Tasman flights of the *Southern Cross* and as *Sydney Morning Herald* representative on the Gull flights had trained him well for the somewhat unpredictable experiences which had to be envisaged in any realistic view of this venture with the flying boat *Guba*.

One of the terms of the charter was that Richard Archbold and his crew should be in *Guba* on this flight, and this was a term I was happy to accept. I was especially pleased to have Captain Yancey as navigator, since I had confidence in his work and could see that in working the aircraft, and in any decisions I might have to make involving the navigation, the personal contact would be a happy one. Moreover we were able to arrange collecting rights for Archbold at each of the islands to be visited.

We left Sydney on June 3 for a nonstop flight of twenty-five hundred miles across north Australia to Port Hedland, which I had selected as the best departure base for the Indian Ocean. In a night brilliant with stars *Guba* rode the easterly wind around the top of the great winter high pressure over Australia, and she brought us in to Port Hedland in the freshness of early morning.

In the late afternoon of the same day we were airborne for Cocos Island, the little atoll which is now the first westbound base on the trans–Indian Ocean air service. The tactics on this stage of the flight were a daylight take-off, for clear vision in the rather restricted waters of Port Hedland, a night flight for stars to navigate the aircraft, and an early morning arrival at Cocos for vision to pick up and alight at this small and lonely objective in the ocean.

As *Guba* climbed away steadily into the western light of the setting sun the way ahead to Cocos Island seemed well established in the sky. Climb to ten thousand for head winds to reduce our groundspeed for arrival after dawn: drift sights in the last of the evening light to discover the wind force and direction and thus to check our early track and speed. Then stars and drift sights through the night to keep her moving for Cocos. A good star fix in the last of the night to establish our position for the run in to the island. That was a simple, effective plan. But there was in it one unknown which had to be accepted. There was virtually no weather information for the route of the flight, though reports from the cable station at Cocos indicated only broken cloud and reasonable conditions at the island.

As the day faded out over the ocean horizon ahead and the brighter stars came into the heavens, that unexplainable union of all life within the aircraft had settled into the cabin of the *Guba*. Yancey had her settled down on a good track, the motors were spinning with the perfect rhythm of space above an ocean, and the night was clear ahead. For four hours we flew in these ideal conditions, till far ahead the lower stars began to disappear.

As we flew on, holding the course for Cocos, a black screen of cloud rose in the sky, gradually cutting out more of the stars, and finally covering all the heavens ahead and above the aircraft.

Soon the wing began to take the shocks of turbulent air and the sound of rain blew harshly on the screen of the pilot's cabin. The peace and tranquillity of starlit space had gone, and we flew in a turbulent ocean of air, close and eerie with the enveloping darkness of cloud. It was unpleasant and somewhat sinister, but it was a good time to have this weather, with plenty of distance ahead and time for it to open into the clear before the end of the night, and for the vital star position before approach to the island.

We climbed her to twelve, seeking to clear the tops, but there she was still in cloud and rain: so we let her bounce along resigned to the weather, and waited for her to break out to the westward of the front.

We had run into the weather at about ten o'clock, time at the aircraft; and by midnight, instead of the improvement or clearing we would have expected with ordinary frontal activity, the aircraft was flying in torrential rain with violent turbulence and frequent glaring flashes of lightning eerily illuminating the ocean of cloud in which we were immersed. I begn to see in this thing the character of a much more widespread disturbance, with possibly embarrassing consequences for the navigation.

A navigator is not a magician. He is a tradesman, and to some extent an artist, who needs both tools and materials for his work. Without the stars for his sextant, and without sight of the ocean for his drift-sight flares, Yancey had no materials, and, without even forecast winds as a basis for reckoning up his compass course, his navigation in the circumstances had to be based upon the best guess he could make: and it would have been sheer luck if this guess had been good enough to hit off Cocos Island.

By about 2 A.M., with no improvement in visibility at twelve thousand feet, we began to push the *Guba* up and finally had her leveled off at fifteen. They only result of this search for the tops was worse turbulence and continuous cloud and rain. With the aircraft scraping at her practical ceiling, in freezing conditions, with the occasional menace of hail tearing at her, we finally had to abandon the attempt to reach to tops of the cloud. Easing the throttles to reduce speed in the surging ocean of air, we sank her down for the more acceptable flight conditions at lower altitude.

We leveled off again at ten thousand feet, held her there at low cruising power to ease the stresses of flight, and staggered on in the most unpromising conditions.

To create a diversion I wrote, from an appreciative memory, a verse which explained everything; and passed it to Lon Yancey.

> We are those fools who could not rest
> In the dull earth we left behind,
> But burned with passion for the West
> And drank strange frenzy from its wind.
> The world where wise men live at ease
> Fades from our unregretful eyes
> As blind across uncharted seas
> We stagger on our enterprise.

He gave me a wry and understanding smile. So—we kept in touch on the navigation; but we both knew there was nothing we could do about it, till we could see the stars and cross the position lines from them. In the back of my mind I kept a check on the fuel remaining.

As the night progressed towards dawn and we still flew in cloud, though with decreased turbulence, Jack and I had one of those short but conclusive conferences which usually come up at some stage of flight.

"How are we going?" he asked.

We were back in the freight compartment down in the hull, where any frank discussion would not reach the crew. I knew what he really wanted to know—where was the let-out from the situation if things went wrong with Cocos?

"We know where we are, Jack, but only within a hundred miles. Until it clears there is no way to get a position. The airplane's in one piece, the motors are O.K. We have fuel for Cocos plus one hour for search on predicted flight time—plus fuel to reach Batavia. I won't let the search go on beyond fuel for Batavia."

"How about the radio; this fancy radio guide thing? Do you think that will work?"

"Probably not: I'm not relying on it."

Back in this hull compartment of the *Guba* there was a somewhat sinister atmosphere. It had none of the frank violence of affairs up front, where weather and the night lived with the airplane. Down there was just the suggestion of safe isolation from the facts of life, but not enough to avoid being vaguely and unhappily reminded of their violence. The whole thing was rather a nightmare, there in the freezing cold, shadowy hull compartment.

Jack was soon busy with his typewriter, composing a press message for Canberra, but I was glad to go up front again in frank contact with the elements. I said nothing about my emergency plans to the crew. I wanted their minds set on the exclusive objective of Cocos.

An hour before dawn we attempted again to top the cloud, for that vital fix from the stars; but again at the practical ceiling of the *Guba* there was nothing but the opaque mists of darkness enveloping the airplane. We held on there, in extreme discomfort from below zero temperature and lack of oxygen, till the first sign of dawn came into the cloud, in which there was not a single break.

In less than two hours Cocos would be due, so the tactics of approach would have to be changed entirely. There was now no possibility of a position from the stars and only a remote chance of a single line from the sun which, should it appear in a break, might give us an idea of our distance run, but no idea of the aircraft's direction for Cocos.

We had in fact encountered a similar freak meteorological

situation to that which put Charles Ulm in the sea somewhere off the Hawaiian Islands after his navigator had run
through a sightless night without stars to a dawn of rain and
cloud.

Now the only chance of finding Cocos was to descend
below the cloud base and fly from an estimated position on a
course for the island. If there was no island at the end of the
time allowance, then we would go into a formal search pattern to box the island.

Guba came rumbling down the height, bouncing on the
rough air through fifteen thousand feet of cloud.

The cloud was hanging low to the sea, so that when we
broke out through the base we were not more than five hundred feet from the surface. There we found ourselves surrounded by a gray and unwelcoming world: a restless, heaving ocean staring up at us coldly without recognition of our
existence. Ahead and around us were heavy showers of rain
with visibility varying from zero to perhaps five miles. It was
a poor prospect for finding an atoll island from a position
necessarily estimated from few facts and many assumptions.

The only thing to do was to fly the course from the estimated position. To have been lured into flying for every
shadow that looked like an atoll in the distance would only
have led to further confusion. The temptation to do this was
strong when behind us there was so much assumption and the
shadow seemed so like an island; but it had to be resisted,
and the tactics kept upon the orderly basis of a plan. *Guba*,
flying at three hundred feet, kept her course and the minutes
of the time allowance for Cocos passed as she flew steadily
on over this bleak, impersonal ocean.

The radio officer now was in clear contact with the Cocos
station for communication, but all efforts to find its direction
with *Guba*'s radio guide equipment produced negative results.
The two needles on the dial, which were supposed to rise up,
cross on a bearing, and thus indicate the direction of the aircraft's head in relation to that of the ground station, certainly
showed signs of life; but they waved vaguely over the dial in
a sightless manner, giving us no real indication of direction. I
watched these things with little confidence in their tempera-
mental behavior, and very soon privately made up my mind

that to rely upon them at all as a means of finding the island would only be wishful thinking. To avoid damping the enthusiasm of the radio officer, however, and other members of the crew who secretly had no belief in any form of navigation but radio, and were inwardly relying solely on this thing, I said nothing of my own convictions but moved in a deliberately unhurried way back to Lon Yancey and his chart table where some practical action could be originated from the little reliable information we had.

Yancey, a first class and conscientious navigator, was showing the signs of his ordeal through the night.

"We'll go after this island, Lon, right to the margin of fuel we need for Batavia, but not beyond it. There's a good chance we'll find Cocos in the time we have; but if we don't, it's better to end up in Batavia than in the drink. How long now to your latest E.T.A.?"

Yancey stepped off the estimated distance on the chart, picked up his slide rule, set it to speed and distance, and came up with the time."

"Twenty-three minutes."

We both looked at our watches. That put it at 0104 G.M.T.—7:34 A.M. at Cocos Island.

There was nothing more to do now but watch, and wait for the island: but we came down just above sea level to set the altimeter for surface pressure; then climbed her again to three hundred feet, holding the course on the compass.

Some of the showers were so heavy that visibility ahead in the rain was zero; so, flying blind, on instruments, at this low altitude, we needed an accurate altimeter setting to give an exact height reading and keep us out of the sea.

The pressure was well below normal for the southeast trades weather we had expected in this region, and though wind on the surface was now from east with a shade of north in it, the altimeter setting was consistent with the freak weather we had encountered for nearly a thousand miles. There was little talk in the aircraft. In fact I sensed a tension in the gaps between personalities, where views were naturally limited to their individual experience and functions in the aircraft. I fancied that if we missed the island the navigator would be held to blame. But I had only admiration for the way in which Yancey was dealing with circumstances which made accurate navigation impossible.

Signal strength continued to build up in the radio, and the gesticulations of the pointers in the radio guide became more frenzied as we flew on the course; but they were the desperate antics of a lost instrument and gave no indication of direction to the station. To placate the crew members who were relying on this thing however, and to give them something to concentrate on so that Yancey would not be harassed by comment, I pretended to believe in this panic-stricken instrument and spent some of the remaining time appearing to be interested in it.

Twice in those last minutes till the E.T.A. atoll islands came up dramatically out of the rain, to lie apparently clearly on the misty sea horizon; but each time, as we approached, the faint yellowish stain of the expected atoll dissolved into lights of the morning where the sun was trying to break through the cloud.

And so the time came, and passed; and only the gray sea looked up at us, always without recognition; and above us the other ocean, of cloud, hung down from the heavens, leaving the aircraft a lone and faraway thing flying suddenly without purpose in the shallow gray air between cloud and sea.

Some definite action had to be taken now, to keep a balanced reaction within the aircraft and to start upon the search procedure. I could not help hearing the very human comment from the pilot's cabin, "Two God-damn navigators, and they can't find Cocos Island!"

There were definite signs of breaks in the cloud, though the sun had not yet come through; so I suggested to Yancey that, instead of going into the search pattern from a virtually unknown position we climb for a complete break so that he could take the sun and locate the aircraft on a single position line. We could then run down the distance from this line to the island, turn down the line and if we then did not sight Cocos, try flying the reciprocal. I checked the times involved against fuel remaining and found that we could adopt this procedure and still have time for a short conventional search and then reach Batavia.

Yancey agreed and I could see that despite his two nights

in the air without rest, he was again alive and inspired now that there was a prospect of locating us on that position line by normal navigation methods.

As *Guba* again ate into the height, up through the sightless ocean of cloud, I thought of the daylight moon. If we could cross the moon with the sun we'd have a position. But when I went to the Almanac and the tables this hope was crushed. The moon would be on a bearing near the reciprocal to that of the sun: so it would tell us no more than the sun's single line. But that line was tangible; something to give us some facts from which we would again set out for the elusive atoll.

At about eight thousand feet the breaks began to show. We were flying through cumulus tops, really bouncing and shaking the wing in the vicious stabbing blows of turbulent air. But the sun was coming through and with another five hundred feet of height we flew in clear valleys of air among gigantic cumulus castles with the sun showing through a thin overcast high above our flight level.

This was the best we could do. Lon Yancey took the sun with only the finest shade on the sextant, and we had the information we needed for the position line. As he worked the sight, *Guba* again sank down through the cloud and finally came out through the base, now at a thousand feet.

Through this climb and descent we had flown on a northwesterly course, deliberately to overfly the island according to the estimated speed which the aircraft had been making good over the sea.

Lon Yancey's sun line put us northeast of Cocos, so we turned southwest and flew the distance between the line and its parallel through the island.

Then we turned southeast down the line; and waited, watching for the island to come up ahead. At this critical time we ran into some very bad rain squalls, where horizontal visibility was zero and we could have run by the island half a mile away without seeing a sign of the sea breaking upon its reefs. In these squalls there was again a strong temptation to divert from the course and search through the rain in the hope that we might fly vertically over the island; but this was a trap and we had to hold on and complete the plan for flying the position line before going into a systematic search to box the island.

All these gyrations with the aircraft caused increasing anxiety and suspicion among those crew members who were not familiar with the navigation. Tension built up in the aircraft as we flew these apparently meaningless courses, and I could see that, now it was obvious to all that we could not expect any navigation aid from the radio, some hopeful clarification of the situation would have to be given to the crew. My attempts to do this were not very successful and I was left in no doubt that Yancey's and my stock as navigators was very low. So I abandoned any further attempts at these personal considerations and concentrated upon helping Yancey in his luckless predicament. Throughout these human actions and reactions the Indian Ocean just stared back at us, revealing nothing.

We ran down the bearing of the position line till the time allowance ran out, and just at that moment, through a break in the rain which lifted the visibility to about five miles, a light patch showed up on the ocean. It's edges indicated the break of surf upon a coral reef.

We turned and flew for this clear indication on the horizon. But again we were deceived by a light effect upon the ocean. As we approached this most tangible image of the Cocos atoll, it melted away in light and shade, leaving only the heaving surface of the ocean still below the aircraft.

Again the *Guba* turned back to intersect the track and straightened up to fly the reciprocal of the position line. As though to impress upon us again the deep and sightless solitude of the great ocean of cloud which pressed down upon us, torrential rain enveloped the aircraft and denied to us the existence of any island.

A check on fuel and distance to Java gave us another thirty-five minutes of search for the island, before I would have to make and effect a decision. Flight up the line revealed nothing. I felt instinctively that we had passed Cocos Island in one of the squalls, but this could have happened at any time when visibility was low; so we were left now with only the search pattern as a last resort to find the island.

At the end of the fuel allowance the search had revealed

nothing; and now the decision would have to be made. I was greatly tempted to continue with the search, because the chances of finding the island with all remaining fuel committed for this purpose were good; but even with the most thorough search pattern of the area we could well pass the island a few hundred yards off in heavy rain, as I believed we had already done. No search could guarantee us passage directly over the island, and no track of the aircraft could otherwise guarantee us vision to see it. Descent in the sea at the end of the fuel might have lost us the aircraft, for a flying boat is vulnerable to heavy seas.

To run for Batavia shrieked at me as a dull and unadventurous procedure; but behind this unattractive view was the purpose of the flight—to cross, for the first time, the Indian Ocean, and to select and survey the island bases on the route. I had gone round the world to search for an aircraft which had the range to take care of such an emergency in which we now found ourselves. We had the means to fly again for Cocos, on a sound flight plan from Batavia. But I looked out over the ocean and felt the attraction of continued flight drawing me away from the aeronautical reasoning, which was clear. There was no doubt in my mind about the right decision; only the voice of the island calling me back.

For a few minutes I listened to this voice. It was attractive and had the appeal which needs no reasoning. Then I thought of the long way which had led to the Indian Ocean flight, of the governments which had entrusted us with the project, and of the awful anticlimax which failure would produce. It was within my decision to avoid the possibility of failure in these circumstances. That was the end of it. I turned to Yancey, who was bending over his chart table, checking back, seeking still a way to the island.

"How about it, Lon? We still have fuel for Batavia plus an hour twenty to dry tanks. Best we alter course now, have a good sleep tonight, check the met for Cocos and come on out tomorrow if this stuff has cleared."

Yancey only needed the firm suggestion. "Yes. I imagine that's best." And he added with a smile, "All the way down the Pacific from San Diego! Those stars. They were just sticking out like lampposts."

"Well, Lon, we'll make it tomorrow all right. Could you give me a course now—for Sunda Strait?"

Yancey went back to his chart, laid down a course from an assumed position, and gave me the little slip of paper with the compass course. In a few minutes we were headed on the new course, and *Guba* settled into her stride below the cloud base.

I went back into the hull and laid out the situation to Archbold. He was perfectly reasonable and accepted the circumstances which had led to the need for the change of course and objective. Jack needed no explanation. He looked down at the still gray and rain-swept ocean and remarked, "Looks very wet down there."

Then we got out a signal for Canberra, setting out the whole story. I could see the headlines—"*Guba* misses Cocos Island," and all the rest of it. It was grim, but it was part of the pattern of the years.

An hour after we had set course for Sunda Strait the cloud began to break, and soon we were able to climb for the tops which the aircraft cleared at eight thousand feet. Early in the afternoon Lon Yancey got a line from the sun, abeam to the northwest; and got the aircraft lined up for the Strait.

In midafternoon we came up with Java Head, at the eastern side, and, passing in by the mountain of Buitenzorg, we alighted in the harbor at Batavia. To wait for a better approach to Cocos we stayed over two nights at Batavia, and in the early morning of the second day *Guba* was airborne for the island.

We flew over Christmas Island and there set course for Cocos, 550 miles to the westward. Though the forecast had been favorable it was not long after Christmas Island that the cloud began to shut in again. We climbed for the tops but were unable to reach them.

Again we could not use the sun for navigation.

I had a good deal of faith in an accurate landfall by using the drift sight, and this was in fact the only method left to us. We descended to below the cloud, in similar conditions to our experience from Port Hedland and, taking frequent drift sights on the disturbed surface of the ocean, kept the compass course corrected meticulously for leeway on the wind.

Half an hour before the E.T.A. at the island we tried the radio for direction. Again it exhibited signs of life but no direction. As the island was coming in, however, it began to settle down and, greatly to my relief, eventually showed the Cocos station dead ahead. Without our altering the course Lon Yancey had given us, in a few minutes the unmistakable outline of the atoll came up dead ahead and, right on his E.T.A., we came in over the Cocos lagoon.

We stayed nearly a week at the island, making an extensive survey of bases for flying boats and land-based aircraft. I had mixed feelings about this work. I knew that the idyllic peace of the island and its happy people must some day have the operations of aircraft imposed upon them. That would be the end of their world. In the overall benefits of air communication between the continents the islands would suffer, and Cocos would be a victim.

Guba left Cocos the evening of June 14 for Diego Garcia, the southernmost island of the Chagos group. Flying under the heavens of a brilliant starlit night we came to Diego Garcia in the dawn. H.M.S. *Manchester*, a cruiser of the Royal Navy, was there to greet us.

After a week of local flights and surface surveys we set out for the island of Mahé in the Seychelles, and there again covered all areas in this region for future flying operations.

The last stage of the Indian Ocean crossing took us in to Mombasa in Kenya, the intended port of entry for the reserve air route.

Back in Australia I wrote up my report and handed it to the Prime Minister in Canberra the day before war was declared on September 3, 1939.

It happened to be good timing. The report was immediately used in the selection of bases needed in the Indian Ocean by the Royal Navy and the R.A.F. for operations against German raiders.

Later, when Japan launched the Pacific war, Japanese submarines operated extensively in the Indian Ocean and antisubmarine operations were stepped up with aircraft flying from the island bases.

When the England–Australia air route was sealed off by

Japanese occupation of the whole area between India and Australia, air communication was maintained with the United Kingdom by aircraft operating through Cocos Island between Colombo and Perth.

After the war His Majesty's Stationery Office publication, *Wings of the Phoenix*, referred to the Indian Ocean operations:

> The irregular chain of tropical islands which extend from Madagascar to Ceylon were mobilized as ports of call for the flying-boats; the Maldives, Diego Garcia, and the Seychelles, etc.
> The coral reefs at the chain of bases, low green islets sheltering lagoons, the palm-covered Seychelles, and blue anchorages that were calm for flying-boats, composed the most fantastic as well as the largest area of operations of the war.
> Catalinas from the remote bases flew hundreds of hours to maintain contact with lifeboats packed with survivors and to guide rescue vessels to them. In all, the flying-boats of 222 Group were responsible for saving more than 1000 lives in the Indian Ocean, a figure which alone is witness to the far-sightedness of those who planned the island bases.

After the war the first air service across the Indian Ocean was inaugurated by K.L.M. in 1949, which operated more than one hundred flights between Batavia (now Djakarta) and East Africa; and today regular services joining the two continents are operated by both Qantas and South African Airways.

Chapter 11
North Atlantic, 1942—1944

WHEN war came in September 1939, my reaction was quite different from that of 1914, when my one fear was that I would not be old enough, soon enough to be in it. This time I felt hostile about the whole thing. I could not escape the thought that we were being drawn into the conventional European war of opposing international political and commercial interests; that because of a lot of comic-opera characters prancing about in fancy uniforms in Europe, we, who had established a clean, uncomplicated life on a continent ten thousand miles away, were being sucked into the whirlpool.

Probably my attitude was colored to some extent by the fact that I had recently married, and though that in no way affected my plans in the air, I really wasn't in the right frame of mind to be thoroughly disorganized by a European war. More than anything, the thought of surrendering my personal freedom of action to the whims of the war machine really alarmed and horrified me. I could see myself at the numerical age of forty-three being put in a chair in some drab and ineffectual service occupation.

With one small new life already on the way for us, my

wife and I took a really good look at this war before making any decision about being involved in it. One alternative was to scrap the whole thing, get ourselves a good sheep property in a pleasant region not too far from the coast, and let the sheep make the war contribution for us with their wool. We did in fact go as far as to look at a property. And one afternoon in the warm sun of a wonderful September day we sat in the grass with a picnic lunch, leaning against a wire-netting fence, looking out over the paddocks to the blue hills of Australia and soaking up the timeless tranquillity of the surroundings. Perhaps it was the very perfection of the country, and the treasured life it represented for us, which made us realize that there was no decision to be made. Silently we both understood that what was at stake was the very life to which we then were tempted to retreat. We knew that we could not live with peace in our hearts if I were to walk away from this war. We turned our backs on the contentedly grazing sheep and the far blue hills, and returned to our car.

Back in Sydney I tried to think out my approach to this thing. I had been a fighter pilot in 1917. Why not now, in 1939? Physically, I knew I could do it; but I also knew that the system would not agree. Yet I had to go through the formality of trying; so I wrote to a friend in high places in London to see what could be done. The inevitable reply came back, as painlessly as possible turning me down because of age. Had this been my exclusive objective, I would have gone to London and perhaps somehow wangled my age and still got in. But the ocean flights over the Pacific and Tasman region had already committed me spiritually to the lure of flight over this ocean and its islands. German raiders were already operating there, and it seemed obvious that eventually the full-scale war would spread to the Pacific. Very little was known of the Pacific Islands from the aspect of air-base requirements. I believed that we would need to know as much as possible about the islands. The quick and effective way to gain this knowledge was with a flying boat. Experience with the *Guba* had proven to me that the P.B.Y. (later known as the Catalina) was the best aircraft for this work. So I drew up a detailed proposal and submitted it to Mr. J. B. Fairbairn, Minister for Air. Briefly, my proposal outlined

a plan to use a P.B.Y. for island surveys of the whole South
Pacific region, and from these surveys to compile an *Air Pilot*
of the Pacific which could be used for future operations a
as a basis for air-base construction. I also proposed that
the course of the flights we should look out for German rai
ers and minelayers operating in the South Pacific. When
cated, their positions could be signaled by radio and ap
propriate action taken to hunt down and destroy them.

The proposal was well received and was in fact approved
by the Minister. I saw ahead the most satisfying and effectiv
war service I could possibly imagine. To be free in the Pacific, with this clear-cut and valuable purpose, was very won
derful indeed. But it was not to be. Jim Fairbairn was killed
when the R.A.A.F. Lockheed Hudson bringing him to Canberra stalled and dived into the ground upon making its final
approach to the airport. My appointment with him, to ar
range details for the P.B.Y. operation, had been scheduled
for the following day. The whole project now became submerged under a smoke screen of opposition and I found my
self back on the deplorable level of seeking interviews in the
drafty passages of Parliament House, Canberra. It was a losing battle, because the project had lost the one significant
minister who combined the authority, experience, vision and
drive to force it successfully through the ranks of its oppo-
nents.

With the inevitable outbreak of war in the Pacific, I again
attempted to reinstate the Catalina exploratory flights to compile the *Air Pilot* of the South Pacific. This was now more
than urgent. But it soon became obvious that the initiative for
action had passed to the United States, and it was not long
before Admiral Richard Byrd undertook a survey of bases in
the Pacific. From this and subsequent American surveys a
splendid *Air Pilot* was compiled and it formed the reference
basis for air operations in the Pacific, including those of the
Royal Australian Air Force. I realized now that Australia
was a dead end for the sort of work which I was competent
to undertake and wanted to do. While these long negotiations
were proceeding, I had kept in the air by joining with Qantas
in the flying of Catalina flying boats from Honolulu to Aus
tralia for the Royal Australian Air Force.

NORTH ATLANTIC, 1942–1944

But now that these aircraft had been delivered, I decided to move out and try to join 45 Atlantic Transport Group of the Royal Air Force, based at Montreal, Canada, which was responsible for flying aircraft built in North America across the Atlantic to the United Kingdom for service with the R.A.F.

I was about to leave for Canada when, early in January 1942, soon after the Japanese had launched the Pacific war with their attack on Pearl Harbor and were rolling back the pitifully inadequate American, British, and Dutch defenses in the far western Pacific, I received a call from my friend Colonel Wym Versteegh of K.N.I.L.M., asking me if I would meet him in Batavia to discuss an urgent flight to the United States. A few hours later I was aboard the K.N.I.L.M. Lockheed flying to the Indies. In Batavia we met in the sophisticated calm of the Harmonie Club, and over the traditional series of Bols he told me the situation, and the urgent need of the Netherlands Indies.

Singapore had fallen. The two great British ships, *Repulse* and *Prince of Wales*, had been sunk with disconcerting ease by a handful of enemy aircraft. What little structure of defense existed beyond the Indies had been overrun by the highly organized Japanese forces according to a long established plan; and Java, the last frontier before Australia, was obviously next on the list. Japanese aircraft were already ranging free over Sumatra and the potential situation of Batavia, the capital of Bandoeng, and all the Indies was desperate. Dr. van Mook, the Lieutenant Governor, had made the snap decision to fly to Washington to seek immediate U.S. aid for the defense of the Indies. Would I make this flight with a Dutch Navy Catalina?

Wym Versteegh's request, and the whole character and purpose of this flight appealed to me immediately. Not only would I be helping my friends in the Indies, but if Dr. van Mook's mission was successful it would also help in the defense of Australia against invasion. Beyond these basic influences was the appeal of the flight itself. Back in the Cat, far from earth, in the deep, starlit nights of the Pacific: Hawaii in the dawn. It was irresistible, and our agreement was quickly sealed. Looking back now I have a sense of nostalgia when I think of our conversation and the very few words which sealed the contract for this important flight.

The air over the Pacific was strange and lonely at this time. There was an unnatural emptiness since the regular civilian airline service had been withdrawn from the threat of enemy action. The Japanese had penetrated eastward to the Gilbert Islands and the vital base of Canton Island stood precariously alone before the flow of U.S. aircraft to the South Pacific began to pass that way. With Canton Island keeping complete radio silence as we flew Dr. van Mook and his party for this necessary refueling base on the way from Fiji for Honolulu, we were not completely sure that it was still in American hands. And when we sighted the island, with several navy vessels lying off the reef, the possibility that they were Japanese had to be considered. We crept up furtively in the Cat, keeping very close below a scattered cloud base, until one of the Dutch Navy officers in the crew was able to identify these as U.S. Navy ships.

We left Canton Island late in the afternoon for the night flight through a sky empty of aircraft in the no man's land of the air track to Honolulu. Always at night over the ocean there is a sense of complete detachment from the earth. But this night particularly, with the fate of the whole region south from Honolulu undecided and in suspense, I felt that our Catalina flying under the stars was very much our world alone.

We sighted the mountains of Hawaii in the early morning, and landed Dr. van Mook and his party at Honolulu for travel on to Washington by the airlines.

After another, similar flight for the Netherlands East Indies and a Pacific air base mission to Washington and London, I joined 45 Atlantic Transport Group and soon felt very much at home in the atmosphere of Dorval and the characters and aircraft who inhabited this transatlantic base at Montreal. This was to be precision flying, with new types of aircraft, in a region of the world where I had never flown before.

Whatever the experience of a pilot when he joined this unit of the R.A.F., he started his service by going to school. Just as I had learned a new kind of flying when I joined Australian National Airways, so did I here at Dorval, with its then modern, high speed, high landing-speed aircraft, reliance on really accurate radio range in low ceiling, bad visibility conditions, and many new technical features in a wide range of airplanes. It was fascinating, and also a challenge. Typically British and R.A.F., there was an absolute minimum of fuss and superfluous academic instruction. For every type of aircraft we were taught exactly what mattered for practical purposes: no more and no less. But, also typical, the very highest standards were expected of us and there was no place whatsoever for "line shooting" or inefficiency. It was an entirely satisfactory life, of the utmost simplicity. I lived in an attic room overlooking the St. Lawrence River in a little place called the Pine Beach Hotel.

My first aircraft was a Liberator. I met my crew in the briefing room before departure. Johnny Rayner, who also lived at the Pine Beach Hotel, was my first officer. The others I had never seen before. They had been allocated to my aircraft by Squadron Leader Coristine, the crew assignments officer. Ed Coristine was one of the world's great diplomats. He was entirely responsible for the crew assignments which, to some types of aircraft, were not at all popular; but I don't think any of even the toughest Dorval characters ever questioned an assignment. One type of aircraft had too short a range of carburetor air temperature control and some of these were lost because the carburetors iced up and the engines stopped. Another had a defect in the exhaust manifold which sometimes cracked and allowed the flame to come back onto the oil tank and set fire to the aircraft. Still another, a particularly notorious one, was lost in numbers without trace till one exploded in the hangar and a type fault was found in the hydraulic accumulator.

The Liberator, however, was a popular aircraft which could fly over the weather, was fast, and could easily do Gander to Prestwick nonstop without going in to Greenland or Iceland for fuel. It was always outrageously overloaded, several tons above its designed gross weight. Furthermore, it had a very fine-sectioned wing which used to flex in alarming fashion in turbulent air. It frankly frightened me, and I could not ignore the thought that in very turbulent cloud a wing might fail: in fact several Liberators did disappear at night in bad

weather on the South Atlantic crossing. In San Diego, California, I later met a man who had been intimately concerned
with the design of the Liberator wing. Pinning him down, I
explained that I had flown this airplane, very overloaded, and
had not been happy to see the wing flex so much in turbulent
air; so would he please explain to me just how it stayed on.
He looked at me with a cynical smile and in a soft but significant drawl remarked, "That's somethin' ah've ben tryin' to
figure out maself."

I wished I had never met him.

But I went out of Dorval that morning with Liberator B.Z.
873 on a new adventure. The weather was overcast with rain,
the cloud base at about five hundred feet. The air was fine
and sharp and the light wind had a strange whispering call
from the icelands far to the north. Instead of warm, colored
islands beckoning from over the horizon of a blue sea there
was something entirely new: a strange open clearness, white,
untouched and infinitely pure with shimmering lights in the
sky, calling from the overcast.

There was an urgent, eager note in the motors as I ran
them up; a quiet confidence in the aircraft as she rolled into
the runway to line up for take-off. But there was also a tense-
ness. Everything always seemed to be stressed to the limit in
a screaming cataclysm of sound as this metal monster projected itself down the runway with it occupants commited
very soon after it started to roll.

A voice from a world I had already left came over the
radio from Dorval tower:

"873—cleared for take-off."

Cowl flaps closed. All set to go, from the final check.

I eased the throttles forward and she started to move away;
heavily started to roll her twenty-five tons of weight for speed
to pass the load from wheels to wing. I touched her with an
outboard motor to check the swing as she thrust blindly for
movement in this early stage of the take-off and then gave
her the full five thousand horsepower. She took it, blasting
her way with all the thunder of the skies for speed to release
her from the earth.

I rode with the Liberator through this terrific, screaming
battle of forces. There was little for me to do but feel and

watch and listen; and be ready for immediate action if she failed for an instant to give all the power she had in the battle with the forces of the earth. I felt the balance of the controls, just letting her go, straight, down the center of the runway, but ready—till she wanted the nosewheel off the ground.

A quick glance at the power—2700 and forty-eight inches. It was all there. The engineer's hand hard against the throttles, turbos set for full throttle take-off.

She tightened down, beginning to feel free. I glanced at the airspeed indicator. Eighty-five. Nearly ready for the nosewheel. Now my right hand rested on the top of the tail-trim wheel. I felt her fore and aft balance with my left, on the control column, and stroked the tail-trim back a shade. There was something definite and satisfactory in the feel of the serrated metal of the trimming wheel. I knew it was going to bring balance and harmony to the whole machine as I eased it back to take all load off the control column, and the nosewheel came away.

Now the freedom of the air was coming. She roared with terrific rhythmic sound, lightly on the wheels as the wing took the load, and I saw the end of the runway coming in. Now it was all the air. There could no longer be any compromise with the earth if any of that tight-strung power should fail. She had to fly. I thought only of the air as she thrust for speed, and I knew it was coming. At 120 she was away, flying, finished with the earth. Only the blur of trees came in under the nose. We set her for the air. My right hand went up in signal for the engineer. He rammed forward the undercarriage lever and the wheels began to move up to bury themselves in the wing. Reduce power: to ease that all-out battle of the motors. I held her down, low over the earth to let her build up speed. Speed, again; to pass beyond that critical early dragging through the air. I drew back the throttles till the manifold pressure fell to 45 inches—felt the elevator trim till she was balanced fore and aft at 160, and let her start to climb.

The engineer signalled the undercarriage was locked up. I pulled down the power to forty-two inches and 2450 r.p.m. At five hundred feet she was brushing the bottom of the cloud. "Flaps up" to the engineer. She sank for a moment, and needed the tail-trim again. Then she began to fly. No undercarriage. No flaps to drag at the wing. She was running clean and free. I eased back the throttles to thirty-five inches and touched the flap of the propeller switches till the rev counters came to 2300. That would do her. She was not so heavy out of Dorval for Gander. Johnny's hand went forward to snick off the booster pump switches. I watched out to the sweep of the propellers and touched the individual switches till they all spun true and the bleating went out of the engines. The last of the earth was swept away by cloud as I went onto instruments for the climb.

At eight thousand feet we broke out through the top. Above, it was bright and clear; clean blue sky without a cloud. We held on to 9,000 and leveled her off; eased the power to 2000 and thirty-one inches, with two inches on the turbos to keep them running; let her cool for a couple of minutes, then cut the mixture to auto-lean. She held 165 on the A.S.I. I lined her up on the course and put her on the auto-pilot.

Everything was peaceful in this new world. Overhead the sky was blue. Away to the north it was faintly green, intensely clear on the cloud horizon. Johnny Rayner sat with a whimsical grin on his face, listening to the Presque Isle range. This was new to me. From far away under the cloud I heard the *a* signal coming in, with a touch of background; then ZQZ the double identification signal. It amused me only to find that the thing worked. Down in the Pacific there were no radio ranges at this time; no radio aid at all for some flights. Bred of necessity on the magnetic compass, the drift sight and the sextant, I had not yet completely discarded my suspicion of all radio aids to navigation. The course on the compass was for Gander airport, Newfoundland; and I expected my navigator to take the aircraft there direct without all this aural contact with earth. My interest in the Presque Isle range signals was academic at this stage, and I didn't want to listen to them. I was having a new experience; seeing new air. The sun was warm over the cloud top, striking into the cabin. Shafts of light moved slowly and regularly on the instrument panel as the aircraft swayed slightly on the auto-pilot. I looked over the oil temperatures and pressures, and across to the head temperature gauges. All were normal. The propellers were spinning with perfect rhythm. There was plenty of fuel. I was warm, comfortable and at peace with the world.

Three hours out we passed over the edge of the cloud shelf and saw below the blue misted surface of the sea; away to the north the coastline of Anticosti Island under Labrador, and down by the starboard wing-tip the Magdalen Islands where Cabot Strait led out to the Atlantic. Away ahead in the distance the western hills of Newfoundland were hidden under formations of pink cumulus, built by the sun-warmed land. I held on at nine thousand. The island came in below, drifting in to the sweep of the starboard propellers: strange that here in the cold gray seas of kelp and granite it should have the form of an atoll; of the dream islands whose names are music in the blue seas of the South Pacific. Here it is Grindstone Island by Cabot Strait, a cold name; an island damp with mists of the Atlantic; strong with the smell of surging seas breaking on dark rocks, close below the cloud. Today, there is sunlight and to the north the icelands and Nova Scotia out of sight behind the wing.

My eyes drifted over the engines. No signs of oil leaks. Numbers one and two head temperatures at 195°; three and four at 185°. I checked the cowl flap switches. All were fully closed. Well, they just ran on the cool side. Oil temperatures all at 72°; pressures seventy-nine to eighty-one. Fuel pressure constant at fifteen. Everything normal. I touched the auto-pilot turn control to keep her on the compass course; called up the navigator for the estimated time of arrival at Gander. In a few minutes he passed up a slip, "E.T.A. Gander 13:35."

The cloud over Newfoundland thickened and lay flat upon the eastward land.

The radio operator handed me the latest Gander weather: "Wind S.E. 10; overcast, ceiling 1800; visibility 8 miles." Improving. If it stayed like that we could break through for a comfortable approach. I put on the earphones, switched on the command receiver and tuned in to the range station at Stephenville, the U.S. Army airport on the west coast of Newfoundland.

I heard the identification signal coming in: *Dit da-da-da; da*. JT, JT. Equal signals—and then the continuous note of the "on course" signal, *a-r-r-r-r-r-r*. Sounds from some hidden, unknown world. A name: Stephenville. We are smack on the center of the range leg, still well away from the station. I switched off the radio, discarded the earphones, and watched ahead over the cloud top. A dark line approached and spread open below us, again revealing the sea. Good air below the cloud. Still plenty of time to break through for a contact approach before the land. I disengaged the auto-pilot, eased down the power and let her go, down for the sea.

At 2500 feet she broke through the base. The ocean was calm and gray. Strange to me. I did not know this sea. Far in the distance a dark shadow hung under the cloud base; the western hills of Newfoundland. I picked her up with the throttles to cruising power and trimmed her again for level flight; touched the propeller switches again till the spinning shadows synchronized and a single note rang true from the engines. Down with Earth, she ate into the cold gray air towards the new land.

I was influenced by two thoughts in attempting this contact flight for Gander. By working the valleys and avoiding the hills in cloud it might be possible to get in with a lower ceiling than was permissible for a letdown on the radio range. If, on route, the cloud shut down to cut out safe visual contact, we could still pour on power and climb into the cloud to a safe height, coming in on the final range approach.

But I really wanted to skate low over Newfoundland: discover this new, dark land, holding the airplane in my hands, and feel the thunder of her engines in the hills, roar her over the uplands, and pour her down the valleys.

As we approached the land, mountain walls rose into the cloud, making an impassable barrier to low flight in sight of the ground. I turned her away to the north and we slid in where the land fell away, edging in towards the course for Gander base and holding deeper clear air on the port side as a ready way of escape. The wing spread over a wild land of rolling hills, hard, bare rocks, scrub, and black lakes like holes leading to darkness within the earth. The air was clean and invigorating and the large aircraft felt responsive in my hands; it was smooth, fast movement, flowing over the land.

I turned on the radio altimeter. It surprised me by rising and falling accurately with the contour of the land, coming down to meet the approaching hilltops and rising as she flung back the rocks a few feet below the wing and sailed again serenely over a valley.

I was happy and exhilarated with the sure flight of this Liberator. Something was fulfilled. I worked her round towards Gander, checking occasionally from the map, and got within about twenty miles when the cloud shut down. I saw it in the distance merging in the hills; no crack of light; no possible outlet by a valley: a thick gray shroud, down before us.

I called to Johnny for "auto-rich" on the mixture to the carburetors, snapped up the main propeller switch to bring up the revs and gave her the power with the throttles. The motors snarled and blasted the air with thunder as she raised her nose and bored up into the cloud. In a moment this strange new world was gone and I watched the little airplane on the gyro horizon in the instrument panel before me, and trimmed her for the climb.

I listened for the Gander range signals, checked the identification, and heard a clear N coming in. That checked our position, west of the airport from the map reading. The radio compass was out; U/S* I turned her on to a bisector to cut the southwest leg and held her up in the climb.

At four thousand I leveled her off for initial approach to the station, and called up Gander control to report our position and request the altimeter setting.

The signals changed as she started to cut into the range leg. Behind the *da-dit* of the N the continuous note of the "on course" signal crept in. It built up, singing through the earphones as the N faded out. I let her go on, waiting for the A to come, to prove we had cut through the beam. *A-r-r-r-r-r dit da-a-r r-r dit da-a-r-r-r dit-da.* Through the beam. We had crossed the invisible, audible road to Gander airport, down somewhere under the cloud. I turned her on the gyro ninety degrees to the left and listened for her to cut back, out of the bush, into the road. Soon the on-course signal started to

---

* Unserviceable.

cover the A. I turned back to bracket the leg and get her running true towards the station. Eighty degrees on the compass
held her, just on the right of the beam. I turned down the
volume as it built up towards the station.

Palmer, the radio operator, handed me up the latest Gander weather: "Wind, East 5; overcast, raining; ceiling 400,
with broken cloud at 200." Bad. I double-checked the altimeter setting for the sea-level pressure so that it would correctly
record our height during the descent; and held on to pass
over the cone of the station.

The sound built up; shrieked, cut out, and it passed
immediately into a clear A. Over the station. Invisible below
us was the Gander Range station, three miles northeast of the
airport. I signaled "Gear down" to the engineer, "Rich mixture" to cover later calls for higher power, and turned her
away thirty degrees to starboard to pick up the east leg. *Da-
dit, da-dit, da-a-a-a-e-e-a-a-dt-da,* and we crossed the leg, easing back the power for a rate of descent that would bring us
out at the right height on our return over the station.

Everything was sound. Two thousand feet over the station;
1256 minimum over the airport—the conditions laid down
for instrument approach. I watched the falling height, the
airspeed at 155; and listened. Twenty-two hundred feet. Two
minutes to go. Ease the descent a bit. *Dit-da-e-e-e-e-e-o-o-
o-w-w-uh-h.* The sound suddenly rose, wailed, cut out and a
clear N came in. Over the station, sixteen degrees to port for
the airport leg. About a minute and a half to the beginning
of the southwest runway. Seven hundred and fifty feet of
height to give away to break through at the minimum. I
called to Johnny for twenty degrees of flap to steepen the descent, touched up the propellers to 2300, let the nose go
down, and drew back the power. She sank on down through
the cloud.

A minute to go. Round us must be the hills; straight ahead
and close below, the invisible airport. With us there was
nothing but the rumble of the motors, the flight instruments,
and the sounds in the earphones. I glanced out an instant,
searching down for the earth we knew was there. There was
only the dark closeness of the cloud and rain streaming in,
tearing harshly at the screen. There was tight suspense in the

invisible closeness of the earth as I held the instruments with my eyes and felt their readings through my hands and feet on the controls. Forty seconds to go. Eighteen hundred feet. No sign of the earth. I let her go down, brought up the rate of descent to go quickly to the minimum, then if there still were nothing I could ease her a shade below it approaching the runway.

Fourteen, thirteen, Twelve-fifty. Twenty seconds. Cloud; no sign of the earth.

"Stand by for the gear."

Twelve hundred. Eleven-fifty. Eleven. Nothing. Blind rain and cloud smoking over the nose. A climax reached up, rushing in on us. No future in this.

"Gear up." I pressed forward the throttles and lifted her out. Get clear of the earth. The motors hauled her away, up from the invisible hills. Wait—for the flaps, more height to cover the momentary letdown. Fifteen hundred. "Flaps up." For a moment she was light under me; sank, picked up, and moved away. I steadied the speed on 160. The little airplane flew just above the bar on the artificial horizon. Five hundred feet a minute on the rate of climb.

"Gander Tower from 879. Am proceeding to Stephenville."

"879 from Gander Tower. Climb to seven thousand feet on the southeast leg and await instructions."

"879: Roger."

I followed the instructions from the Gander control and worked her up through the cloud. I didn't want to be delayed. Stephenville would be open for day approach, but uncertain for night. I wanted to get in as soon as possible. In a few minutes he came back.

"879 from Gander Tower. You are cleared to Stephenville. Weather: calm; overcast; ceiling two thousand; visibility eight miles."

"879: thank you."

I pulled her away and straightened up on the course for Stephenville. In forty minutes I tuned in to the Stephenville range and brought her in over the station.

I got a signal through to Gander in the morning for Atlan-

tic clearance out of Stephenville. Walking down to the aircraft I picked up a round, waterworn stone by the side of the
road. It was very smooth and cold; hard with long endurance,
like the hills and the rocks we had flown over. I could see in
it the serene faces of the lakes and the still black ponds, the
waters rushing down through hills of pine and birch trees to
the strong gray sea. In the air on the way to Gander my fingers felt the deep texture of the stone in my pocket. I took it
out and tossed it across the cockpit.

"Catch, Johnny. That's Newfoundland."

We went out of Gander in the evening. She was tight and
heavy, with full Atlantic load. I held full power as she
plunged forward into the darkness off the end of the runway,
and took all we had from the engines till she built up some
speed. The weather was down again. In a few moments even
the coal-black darkness of land was gone and something close
and faintly lighter than the night closed around her as she entered the cloud. There was an impression of a critical situation in the aircraft as she dealt with the heavy load in this
blind struggle for the air above. The world was close before
us in the luminous instruments that picked up a glow from
the fluorescent lights and stared at us securely from the
panel. I was relieved when the wheels were up in the wing,
the flaps drawn back into the trailing edge and the power was
eased from that all-out blasting struggle that left no margin.

I gave her plenty of speed and power, and watched the
flight instruments to keep her closely trimmed, trying to settle
her to a steady climb as she hit the uneven air of the cloud.
Outside there was nothing. Here in the pilot's cabin there was
a strange security, a warmth of life in the instruments, the
rhythmic roar of the motors, and the inhabitants of our
world; but there was tension working for the height before
we could feel the easy swinging stride of cruising flight, out
with the stars above the cloud.

We broke through the surface at nine thousand feet and
held on to eleven-three before again reducing power. I lined
up the auto-pilot and snapped on the switch as each pair of
lights flickered out. As she settled at eleven thousand on the

* Airspeed indicator.

altimeter the speed held steady at 175 on the A.S.I.* I trimmed her there, let her go, and sat back to relax and enjoy the flight.

Now she was set in her own orbit, moving across the universe with the other stars, steady, above the gray mists of space faintly visible below.

Jack Hood, the engineer, handed up mugs of coffee. We needed it now, after the strain of taking the overloaded aircraft away from the earth and up through the turbulent weather. I didn't particularly want to know anything about our position for at least two hours. There was every kind of aid to navigation on the Atlantic. And we knew the weather to be good for this crossing. I wanted the navigator to conserve his energy so that he would be alert in the morning, and give me a reliable, exact position approaching Ireland.

Palmer, the radioman, handed me a note.

"We have a passenger. Sitting on my receiver."

I looked back to the radio cabin. It was a large green grasshopper.

I called to Palmer, "He looks a bit groggy. Try him with a shot of oxygen."

We shot him a whiff from an oxygen line and he sat up, looking a good deal better. He scratched his head with his back leg and appeared very comical, like a horizontal giraffe with front legs in its neck. Palmer got some lettuce for him out of a sandwich box, and we left him to it.

We ran through a light warm front about an hour out and then came out to night so clear and still and black that the airplane seemed frozen motionless in the heavens. We got a star position when necessary, kept radio watch, kept the fuel up to her from the bomb-bay tanks and continuous watch on the compass course, altitude, and engine temperatures and pressures. I didn't sleep. I seldom do in an aircraft, except when traveling in a resigned condition as a passenger. There is always something. Something to watch out of the corner of your eye, to think about and decide what you will do before it comes up; and always something new, however small, to take from the air. When a man reaches the stage where he feels he has learned all there is to know about the air, when he is simply bored, with nothing to do because everything is going well, it is time for him to stay on the ground.

About four o'clock in the morning I turned from an awed contemplation of the northern lights and relieved Johnny on the pilot's watch. The first signs of dawn were in the east. There was only a faint difference in the sky; just enough to give you that new sense of relief that day was coming over. We had been silent a long time.

"It's cracking, Johnny," I called across.

The effect was electrical.

"What's cracking?"

"The dawn, Johnny."

We both lay back convulsed with laughter, speculating on the things that could crack in the night over the Atlantic.

Palmer passed me a message. We stopped laughing. "There's a Fortress in trouble. He's lost an engine and he thinks another is packing up."

I pulled on the earphones and switched on the intercom. "Get his position and course."

What could we do? Drop him our rubber dinghy perhaps, if he got down and were able to show a light. We could stay with him for about three hours, till they got a Cat or a patrol vessel on the way.

"He's sending the S O S. I can't contact him."

"Keep on trying."

"Can't hear him now."

We didn't hear him any more.

Five minutes later I saw a bright light far ahead on the sea. It looked like flames, burning up and dying; flaring again. But it was a ship. She passed right under us, brilliantly lit from end to end, presumably a hospital ship or a neutral. Nobody heard the Fortress again. She just didn't arrive at Prestwick.

Morning came with a sea of cloud below us. The stars had been good. It was still clear above and Craske, the navigator, had the position well fixed. Soon he would take a final distance out, from the sun. That would be all we could take from the sky. There was a big buildup over Ireland and the west of Scotland; with low cloud round the land; so the final approach would be a radio job, probably with a letdown on the Prestwick range. As a further check on our position Palmer had got a radio fix from Iceland and southeast England. It tallied with the star position. We were all set for our approach.

Visibility above the cloud was perfect. We were only two hundred miles out in the Atlantic now, in the region where it was wise to keep watch for the long-range intercepting Focke-Wulf Condors. I sent back instructions for a watch to be kept out the tail by the two passengers who had had the misfortune to travel with us amongst all the military equipment and drafts that whistled in through the gun turret. It would give them the dubious interest inspired by the possible anticlimax of being shot down while enjoying the morning.

About seven o'clock I tuned onto 225 KC and listened for the Derrynacross Range. It was there all right; just a faint N in the "on course" signal persisting through a lot of intermittent crackling, and the faraway squeak of the double identification. The sound of that range was like the lonely cry of a world lost in the depths of the universe. It seemed infinitely far away; calling, in the failing hope that somebody would hear, and come.

As our reckoning told us that the coast of Ireland would be coming in soon, towers of cloud rose up before us and there were great gaps of smoke-blue darkness below. Through one of these I saw a wandering white line and suddenly realized it was the Atlantic surf breaking on the rocky coast of Northern Ireland. Then there was a bright green field and a patch of red earth; the world; people would be living there. Immediately it was wiped away; gone, surely only a picture in my imagination, as we drove surging into the cloud and again there was nothing but ourselves, the roar of the motors, and the instruments in the panel before us. She began to bounce around, stabbing into the rough air and shaking the springy wing in a way that made her feel uncomfortable to me, too mechanical on the auto-pilot. I switched it off and flew her through the weather. I heard her cut through the north leg of the Derrynacross Range, the radio screaming and crackling through the great mountains of cloud and rain that lie over the hills of Northern Ireland. I made a quick mental estimate of time for Prestwick and called for an

E.T.A. from the navigator. We stabbed on through the cloud, the flexing wing shaking the whole structure of the airplane. Quite suddenly the whole sky opened up before us and we flew out into clear air. Coming in under the inboard motor I saw the Mull of Kintyre, and down over the starboard nose the great rock of Ailsa Craig showing through a wisp of low cloud. I eased down the power and descended to minimum height for safe approach. As we came in over the Prestwick station the cloud broke up completely. We let down the gear and straight away got clearance into the airport circuit. We set her up for the landing and, descending, she came around facing up for the runway. Flaps down. She sank in the last few hundred feet, dragged through on the engines. Everything off. The black surface of the runway rushed under her, the wheels took the load from the wing, and our metal monster came to rest.

Chapter 12
Gander Lake

Back at Dorval I found Ed Coristine stacked up with Catalinas for delivery to the U.K., with most of his boat captains already away on the route; so I lined up with a Cat delivery to Largs, at the mouth of the Clyde in Scotland.

Now, in the summer we could leave from Gander Lake for Largs; but in winter, though the lake was not normally frozen over because of some warm underground springs, we could not take off from the water because immediately upon opening the throttles the propeller spray would freeze on the cold airplane, turning it into ice. So for several months of the year the Cat had to fly the long nonstop haul from Bermuda to Largs: about three thousand nautical miles and anything up to 30 hours in the air, with a single crew.

For this Catalina delivery we were on the summer route, from Elizabeth City in North Carolina where we collected the airplane, to Boucherville flying-boat base at Montreal, to Gander Lake, and across the Atlantic to Largs.

I had a very young R.A.F. crew, ex-Coastal Command, with flying officer Bowman as my second pilot. There was something fresh and adventurous about these chaps and it was a happy crew from the start. We went down to Elizabeth City, did a test flight and flew over the Wright Memorial at Kittyhawk.

From Elizabeth City we flew the airways north by New York, Albany, and Burlington, to Boucherville the next day, and out to Gander Lake.

We went out of Gander at dusk the same night. About two hours out I was running the flashlight over the engines when I saw a stream of oil coming back under the port cowling. I watched the ripples of oil streaming like waves on the surface of a wind-blown lake. Such waves in a thin film of oil look like a constant stream leaking from the engine. A cupful spilled on the cowl can make an engine look as though it is bleeding to death. So I was not immediately concerned about the glistening ripples showing in the beam of the flashlight. We had forty gallons in each tank; but it was a warning. The cowl was dry when we left Gander. It was dry all the way up from Elizabeth City and Montreal. Now there was oil on it. I handed over to Bowman and went aft into the blister compartment to watch for oil coming away into the airstream.

It was coming all right. The ripples were ending in little blobs, detaching themselves and flicking away into the night. There was definitely a bad oil leak somewhere.

Ahead of us was fifteen hundred miles of North Atlantic Ocean; ten hours of moonless night before the dawn. The leak could be a loose connection. It could be a cracked line. It could come away suddenly, letting go all the oil from the port motor, leaving us to fly heavily laden on one.

Gander was open; cloud base at 2500 feet. I returned to the pilot's cabin, turned her round, and asked the navigator for a course.

Bowman asked me if he could fly her instead of putting her on the auto-pilot. The air was rough and we passed through broken cloud towering high above us. He was keen to practice flying on the instruments, to do anything to increase his experience of handling the aircraft. I left her to him and as we shortened the distance in towards Newfoundland, put on the earphones and tuned in the Gander range.

Soon we saw ahead the lights on the east coast of Newfoundland as they crept in slowly out of the darkness. Others came, and spread away as we passed in over the land; many more lights of human habitation than I had expected on the coast of this lonely island.

Then, the Gander lights ahead. In a few minutes we were back over the airport, now a pattern of lights in the darkness below. The tower came in with instructions: "Hold at two thousand feet till further instructions. Flares are being laid for you."

We let her go on for the Lake, down over the hill from the land-plane base, and started to circle the area. It was just possible to see the difference between the lake and the hills, where a long, coal-black shadow lay encircled by the eerie darkness of the land.

Down on the edge of the shadow I saw a light moving. Must be one of the boat hands seeing about the flares. Another light pricked the darkness and the two moved jerkily around each other. For a moment I lost all touch with the ground and had to go on instruments to check the attitude of the aircraft. She came on round in the turn and again swept the airport lights into view. They passed by the nose, and we drifted on round to oblivion again.

No sign of activity at the lake. Perhaps there was trouble with the floats that hold the flares, or possibly with the lighting itself.

We circled the base for nearly an hour, held by instructions from Control. I began to feel restive, and called the tower.

"Gander Tower from 925. How long before the flare path will be ready?"

"Nine two five—stand by. Will advise you. Over."

Suddenly a whole snakelike constellation pricked the darkness and began to writhe and twist in uneasy contortions in the starless void of night below. Then with a great effort, it slowly straightened itself out and began to slide across the black hole below us. I held her in a steady turn over the lake now, watching the floating flare path being towed into position. It slowed, stopped, and the star snake seemed to collapse and sag back into an aimless pattern of drifting lights. It made another convulsive effort to straighten out, relaxed, twisted its tail and drifted aimlessly round towards the edge of the black shadow.

The flare situation wasn't working out at all. I decided to land without it and called up the tower.

"The flares are drifting ashore. I am going to land without them. Would you please request the marine base to clear the landing area? What is the height of the lake surface over sea level, and the altimeter setting?"

Actually I was keen to try an instrument landing. Very recently, having in mind such an eventuality as this, I had practiced blind landings in daylight on the river at Boucherville, and was quite astonished to see how well they came off, provided exactly the right airspeed, rate of descent, and directional gyro heading were established and maintained. To deliberately sit there not looking out to land as the surface of the river came up to the airplane had been a horrifying experience at first; but it had worked, and I was intrigued now to use this experience for good reason in the really blind conditions at Gander Lake.

In a few minutes the Tower came back again: "We are checking the lake level for you. The altimeter setting is one zero two zero millibars."

I turned my flashlight on the altimeter, reached forward to the adjusting knob and turned it till the setting was exactly ten-twenty millibars, sea-level pressure.

Down on the lake the lights slowly went out. In a few minutes there was only one. Then that was gone, and there was only the dim black shadow in the night. I leaned her round on the air, waiting to hear the voice which would tell me the level of the lake. Soon it came.

"Nine-two-five. Nine-two-five. The surface of the lake is one hundred and eighty-one feet above sea level. One-eight-one feet. You are cleared to land, at your own responsibility."

I took her over the shadow line that marked the position of the marine base on the shore, took a mental picture of that part of the lake where I wanted her to touch down, and turned away to fly into the southeast, to return on a long straight approach between the hills.

I pressed forward the propeller controls for 2300 revs and

snick up the signal light switches for the engineer up in the tower below the wing. The little squares of light went on in the switch panel before me.

"Floats down."

"Auto-rich."

There was a sudden signal of movement within the aircraft as the float mechanism began to function, and two dark canoes dropped out of the wing tips and lowered themselves down to fly alongside like sharks that swam with us in the night. Then a jerk as they locked down ready to give her stability when she came to rest on the water.

I fed her some power to beat the extra drag of the floats and held her at ninety knots, descending gradually for the southeast bend of the lake. Four miles from the base I swung her round, keeping away from the hills, and straightened her up at 700 feet on the altimeter. Five hundred and nineteen feet to lose in the four miles. One hundred and fifty feet a minute till the altimeter showed the approaching surface. Then break it gradually to fifty and wait for her to go on.

I got her running true up the center of the lake, noted the gyro heading, and went to the instruments, eased up the nose and adjusted the throttles till she held seventy-five knots on the airspeed indicator and 150 feet a minute on the rate of descent; watched the gyro on three-twenty degrees.

The air was dead still, in black darkness. There was not a movement in the aircraft as she stole quietly up between the hills, sinking steadily in for the invisible surface of the lake. I glanced out for a final check on the accuracy of our heading. The hills closed up around us and we seemed to be sinking into some abyss of night, a bottomless hole in space where there were no stars.

Nothing more outside. I went back on the instruments, committed now irrevocably to their accuracy.

The height was going. The long needle of the sensitive altimeter crept down to three hundred feet. I took a slight tension on the wheel to hold the speed down to seventy-five knots. My hand went up for a quick touch of the tail trim for the final balance fore and aft. The gyro was steady on three-twenty.

Only a hundred feet to go. I held her with my left hand,

feeling her down; reached up again for the throttles with my right to bring up the power steadily, breaking the rate of descent to fifty. The needle rose towards the luminous bar of level flight. I drew off a touch of power—seventy-five, fifty, three-twenty. Seventy-five, fifty, three-twenty. . . . Everything was concentrated on the figures in the instruments.

Two hundred feet on the altimeter. Seventy-five—fifty—forty—thirty. Near the surface. I felt the cushioning effect of the air squeezed between wing and water, lifting her, to break the rate of descent. I held her dead steady, not moving anything. The aircraft was set floating on the air in quiet suspense, very close now to the invisible water. I waited for the sound which would tell us the hull was cutting in. Then it came, like the rush of steam escaping under the keel; and I felt the hand of the water stroking the hull, caressing her down. I still held her steady; drew back the throttles and took the thrust from the propellers—steady, without a movement fore and aft while the water slowed her, gently, but firmly; it rushed in under the hull and took her from the air. The wing had finished. Slowing, she rode up on the bow wave and sank back gently into the waters of Gander Lake.

I looked out into the darkness. There was a light away under the starboard wing. We had run some little distance past the base. I swung her out and round with the starboard engine and headed back for the moorings. A spotlight stabbed out across the lake and moved swiftly for the moorings. The speedboat circled the buoy, moved away and held it for us in the beam. It was unnecessary. I would rather have had the surface undisturbed, so that the aircraft could move accurately through the water, pick up the buoy in her own light, and trickle up to it on the drogues. With waves from the speedboat swaying and lurching the aircraft, we picked up the mooring and there was a great disturbance around us; boats milling around stirring up the lake; breaking into the silent wonder of the night with their discordant confusion. One of the coxswains passed me a message. I was wanted at the airport. There would be a station wagon down at the dock. The one thing that mattered was that the oil-leaking Catalina was down and on the mooring. The engineers would be over in the morning. I had no intention of going to the airport. I asked the Newfoundland boatman to telephone Operations from the dock, to tell them I would be sleeping in the aircraft tonight and to have the engineers down at the lake at daybreak. And I privately arranged with him not to allow any boats out to the aircraft for any reason whatever.

I got my crew away for their quarters as quickly as possible, but the second pilot hung back, looking at me as though he wanted something.

"What is it, Bowman?"

"I'll stay aboard, sir. I'd like to. Seems strange our going ashore and you keeping watch in the aircraft."

I'd planned to stay alone and was looking forward to enjoying the deep silence of this primeval lake by myself; but I could hardly refuse him.

"Well, we'll both stay."

We rigged up bunks in the cabin and undressed. I went aft into the tail, lifted the tunnel hatch, and felt down into the water a few inches below. It was cool and refreshing, but I might have been looking from the lighted cabin ten thousand feet down into the darkness waiting for a drift flare. Like the lake from above, the open hatch was just a black hole in space, though the water was only six inches from the swept-up bottom of the hull. I lowered myself down into it and splashed around with my feet. My imagination went back to the low flight over the dark ponds with the Liberator.

I slipped down through the hatch and swam out from under the aircraft. I swam fast through the darkness, turned and came back. It was very eerie in this deep lake. Slipping in under the aircraft, I heaved myself up through the hatch.

We made some hot chocolate on the electric stove, sat out in the open blister compartment and talked. There was infinite peace in the silence of the lake where the earthy smell of Newfoundland drifted on cool air from the pines and the silver birches in the valleys.

Then I lay in bed, lived through that descent again, waited again for that keen, sizzling swish of the water on the keel. The last thing I heard was the music of ripples tinkling against the metal of the hull.

The oil leak was from a loose connection; a new-type

clamp that was supposed not to need locking wire to keep it secure. It had shaken loose and was letting the oil away.

We left Gander late that afternoon. Bowman flew the Catalina by hand most of the night across the Atlantic and I let him land her when we went into Largs in the morning.

But a few weeks later he was killed in a bad "porpoise" landing when the last bounce stove in the nose. He had to sit without touching the controls, still in the second pilot's seat.

Chapter 13

Hurricane at Clipperton Rock, 1944

ONE of the exploratory flights I had planned as a lead to an international air route of the future was the survey of a route joining Australia with the United Kingdom and Western Europe by a direct line across the Pacific to Mexico and, after touching at a main eastern traffic center of the United States, on across the North Atlantic to Europe.

But in 1944 the survey of routes for civil air services was submerged under the more urgent needs of war transport and communications. Since the war in Europe was obviously in its final stages at this time, the diversion of R.A.F. operations to assist in finishing off Japan after the defeat of Germany was now in sight; and a convenient ferry and communications route to the Pacific was being considered. With this in view I had already approached the Commander-in-Chief of Transport Command, Air Chief Marshal Sir Frederick Bowhill, with the proposal that we should explore the Central Pacific line to the Southwest Pacific as a direct outlet from the Atlantic for R.A.F. aircraft to that region. My proposal had received a favorable reaction from the C.-in-C., but there were certain international problems confronting this operation. The Pacific was an American theater of war administered from Washington, D.C. Before any R.A.F. survey flight could go out into this region, U.S. approval would have to be received. Such an operation by the British was viewed in some U.S. quarters as a move designed to gain an advantage on a future civil air route. In point of fact it was completely genuine. The North Pacific route by Hawaii was already overcrowded with aircraft; the R.A.F. was genuinely preparing to throw its full weight into the Pacific war, and a second ferry and communications route was needed. The fact that it had some future significance in civil air transport was quite incidental.

But the problem of U.S. approval had to be overcome and the approach to it had to be through the R.A.F. delegation in Washington. I saw immediately that very strong instructions would have to go from London to the senior officer of this delegation; instructions in the face of which he would feel that not only must he approach Admiral King, but he must do it with the firm intention of getting his approval. But I had already had more than enough experience of diplomatic touchball to know that if we were to get results in this case something more than the conventional approach would be needed.

I was at this point in London and upon the principle, long ago learned in the air, that to get the best results it is usually better to ally the favorable forces to your end than to try to force a passage through the unfavorable ones, I rather amiably allowed myself to be diverted to some R.A.F. officers of the Air Route Planning section. These chaps, fresh from operations, were alert and imaginative, and over a few beers in the local pub under Westminster Bridge they were very soon inspired with enthusiasm for the R.A.F. route to the Pacific. In this tavern, rich with the smell of good ale, we drafted a signal which, with a smile in our hearts, we felt would produce the necessary results in Washington. My fellow conspirators undertook to escort it to the high destination for its signature. There was a brief interlude as we heard the high scream of a buzz bomb approach and pass overhead; then the silence as the engine cut out, and the deep, muffled explosion somewhere in London. I knew at that moment that from London the Pacific was a distant and perhaps intangible region, but these R.A.F. chaps could see it. We drank to the success of our signal, and went our ways.

I reported back to the C.-in-C. at Transport Command Headquarters at Harrow. I purposely did not embarrass him by telling the details of my slightly subversive activities, but appropriately informed him of the impending instructions which I understood were going to Washington. I also suggested to him that my presence with the R.A.F. delegation there might be helpful with technical details of our proposed operations. With perfect understanding, he arranged with Air Vice Marshal Marix, A.O.C., 45 Group, Dorval, to have me sent down to Washington to aid the delegation in this way.

In all this campaign for action on the Pacific route I was greatly helped by my status in 45 Group. I was at that time a junior civilian captain, with one bar on the shoulder strap of my tunic. The normal effects of service rank could hardly apply to me, and the C.-in-C. graciously seemed to accept me for what I was: a pilot and navigator of aircraft with a wartime purpose which I believed could best be applied through service with Transport Command.

I think I understood the localized feeling along the line of conventional approach to our Pacific survey operation, and should perhaps mention it here.

When I had previously come to London, fresh from an Australia threatened by imminent Japanese invasion, I was charged with a sense of importance in effective war operations in the Southwest Pacific. I stayed the first night in a small and intimate hotel off Jermyn Street, and was shown to my room by an elderly hall porter who with considerable dignity apologized for the fact that the whole top had recently been blown off the building by a bomb, except for the lone room which was my accommodation. He might have been excusing the hotel for some trivial and temporary inconvenience like an unserviceable room telephone as he stood in the midst of the most awful shambles of rubble and showed me to my room. I could not help being impressed by his complete refusal to show any signs of the seriousness of a partially destroyed building which might well in the next air raid be written off altogether. I felt rather small actually, in the face of this old man's dignity and his concern only for my personal reaction as a hotel guest.

Late that night some German aircraft came over. I heard the wailing of the sirens and felt the gradual hush come over London. I got up and looked out from the window, over the dark roofs and into the blind heavens. I had a feeling of being a spectator; that I didn't belong in this. This was the old hall porter's world. This great, deeply breathing city was crouching in the darkness, calm but charged with a grim menace to the approaching Germans.

The shells which burst high up in the night round the enemy aircraft caught in the intersection of the beams had little significance. It was the silent, invisible city which impressed me. And I knew why my world in the Pacific seemed so distant to its inhabitants.

To reach Washington in time to ensure that the signal I had originated would receive attention I boarded an R.A.F. aircraft for Prestwick, very nearly got shot down by our own guns when the pilot flew over some prohibited area, and escaped back over the Atlantic in the return ferry. In Washington my reception was cordial enough, but I was soon passed down the line to people who had obviously been deputed to get rid of me. There was, it was said, really nothing I could do, since the whole thing was now on such a high level that it was spoken of only in the most guarded whispers. I had been subjected more than once to this high-level routine which is intended to impress one with its sanctity, but I was not impressed by it and rather cruelly stayed on in Washington to get results. Air Marshal Welsh, however, proved himself to be a first-class diplomat, for he very soon saw Admiral King and had his clearance for the R.A.F. operation in the Pacific.

I hurried to Bermuda where my aircraft, Catalina J. X. 275, was ready on the ramp and the crew standing by for immediate departure at dawn the next morning. But that very night a signal came through from Dorval canceling the flight. I had, after all, been outwitted in Washington. It was now said that this was a matter which could not be decided on any service level and that the personal sanction of President

Roosevelt was needed. I flew up to Montreal to see the A.O.C., and that evening drove up to the Laurentian Mountains to stay with my family, for whom I had recently managed to arrange a passage across the Pacific from Australia. Here we had a lovely little ski shack in the peace of the Canadian woods. I had just arrived when the great news came through by telephone from Dorval. Air Marshal Welsh had got President Roosevelt's approval. In a few hours I was back in Bermuda, and away for Acapulco. Upon finally leaving the Pacific Coast of Mexico I instructed my radio officer not to receive any signals recalling us to base.

And so we flew out to find Clipperton Island, the little atoll on the track for Tahiti, uninhabited for twenty-seven years, and now a forgotten Island.

We picked up the island from a sun position line and found a way in to alight on the coral-studded lagoon. We laid down four hundred gallons of fuel in cylindrical tanks which we moored in the anchorage. Then we flew back to Mexico, refueled to capacity and came again to Clipperton Island, now in a position to leave there with fuel for the three thousand nautical miles flight in unknown winds to Bora Bora.

But we were destined to be marooned at the island for six weeks before the Catalina finally rose again from the lagoon and headed out into the blue for the long run to the Marquesas Islands and Bora Bora.

During this time we converted one of our rubber dinghies to a sailboat, building a leeboard structure to fit over it, and mast and spars from timber found in the ruins of the old settlement; and making a mainsail and jib from some cloth we had brought out from Mexico. With this strange craft we were able to work to windward on the lagoon, as well as to lead and run before the wind, and we made with it a complete survey of the flying-boat area and the coral patches which would have to be blasted and cleared before the base would be suitable for regular operations.

Our compulsory stay at Clipperton Island was caused by some engine trouble which built up into a complicated situation from what had seemed a relatively simple one. During a plug change two spark plugs broke in the cylinders; these had to be removed and some major work done, for which parts had to be brought out by a reserve Catalina standing by in Mexico.

The whole thing developed into a sort of Robinson Crusoe experience in which we really became very detached from the world, living to a great extent on the natural foods of the island. We made spears from old iron rods in the ruins and learned how to spear fish in the reef waters. We climbed the palms to get coconuts in the small grove which existed near our camp on the otherwise treeless coral rim of the island, and we found wild spinach growing in pockets of soil produced by long-decayed vegetation. At one end of the lagoon Clipperton Rock (the *Rocher* Clipperton of the old French chart which we had) stood up seventy feet above its flat surroundings. The rock is the peak of a submarine mountain rising twelve thousand feet from the Pacific sea bed and standing above the surface for only the last seventy feet of its pinnacle. Clipperton Rock, seen from the distance when we sighted it before the island rim was visible, was exactly like the sail outline of a full-rigged ship upon the horizon. White with the deposits of millions of seabirds through the centuries, it has an ethereal appearance quite in keeping with the whole impression of this lonely island. At this rock we found traces of habitation by the lightkeeper who years ago had tended the light, which was still standing on its pinnacle.

In the ruins near our camp we could see much of the story of the last inhabitants written in the still and pathetic remains. In about 1906 a British phosphate company had obtained a concession from Mexico, then recognized as the owner of Clipperton, to work the deposits on the island. A community had settled there, with the Mexican garrison and the families of the phosphate workers. All had gone well, with a supply ship coming every six months and taking off the collected phosphate deposits, till the outbreak of the first world war. After some time, and for uncertain reasons, the supply ship just had not returned. The plight of the people on the island had become desperate for want of food and rampant sickness. Many had died, the rest had become weak and exhausted. After a time only a few men, some women and small children, and the giant Negro lightkeeper remained. This man decided to kill the other, weaker men, enslave the women and live as a kind of king of the island. He had been successful in his original endeavor; but a young woman of obvious spirit and initiative named Tirza Randon had waited for an appropriate moment, quietly taken an axe and smote him very effectively on the head with it. Miraculously the few survivors on Clipperton had been rescued the very next day by the U.S. Navy vessel *Yorktown*, which passing close by the island had seen their signals. After a long dispute about its sovereignty, the island had been awarded to France. We were constantly reminded of these earlier Clipperton settlers as we found various material and personal things lying beneath tangled masses of vines; and although we did so reluctantly, we had to use some of these.

As time went on at Clipperton and the engineers went on with their work, I became increasingly restive about the safety of the aircraft on her comparatively light anchor and cable in the winds which every few days came over the island in the apparently normal cycle of weather. Finally, after one very bad night waiting through the hours of jet-black darkness and driving rain to start the engines if the anchor picked up or the cable went, I made up my mind somehow to lay a mooring; for even if we used the engines, the amount of coral in the lagoon was safely negotiable only in daylight and then with the sun behind us. To be loose in the lagoon at night on the engines would have been a bad experience.

Over in the ruins there were some heavy rail truck chassis; and we had found some old ship's chain under the vines, corroded by rust but still very strong. In the aircraft we had some fairly heavy Manila line. I tried to think out a way we could combine these in a mooring in the lagoon by the camp and have the airplane infallibly fast to it so that we could sleep at night with confidence. After some thought we devised a system to lay this improvised mooring.

We slid the truck chassis down to the edge of the lagoon on long wet planks, and dragged down the chain to the same position. Then we built a timber platform across the two rubber dinghies, which together gave buoyancy for more than two tons. Onto this strange barge we managed to manipulate the heavy steel chassis, again by sliding and levering this heavy equipment across on the strongest timber we could find. We threaded the chain through convenient parts of the trucks, attached a double thickness of Manila to two separate lengths of chain, fastened floats to the line, and we had the mooring afloat for transportation to its site.

The tricky problem was how to get this contraption off the improvised barge without somebody being dragged down with it or hit by it.

We started by paddling out to the mooring site. Then we gradually levered the whole thing across the barge till it was reaching a point of balance. We carefully checked the impending runout of chain and line; everybody got clear of any entanglements, and with a great heave we upended it. There was a terrific roaring crash as it plunged into the water; bubbles seethed to the surface, and the floats alone remained in sight. Very soon we had the aircraft on the mooring.

The base aircraft, 603, was also at the island, having brought us out our spares. These had been fitted to the engines, a general check made all round, and we were ready to attempt the heavy overload take-off for Bora Bora from the critical length of runway available through the coral in the lagoon.

On the mooring, we slept well the night before our projected departure.

I awoke soon after dawn and from my bunk in the blister compartment looked out at the weather: for the wind chiefly to see if it was blowing from south, where we wanted it for the longest run on the lagoon; but the aircraft was swinging with her tail towards the camp, to a fresh breeze from north. It was coming in little explosive puffs that spread out on the water under the windward shore and hurried towards the aircraft, rippling the surface darkly.

Watching this wind, and weighing up the chances of take-off, I was conscious of the sky. It was dark and heavily overcast, with nimbus cloud hiding the tops of cumulus that reached into this high and dismal covering. Below it all, low gray scud was hurrying furtively over the island as though not wanting to be seen. It was moving fast; much faster than the air five hundred feet below it, brushing the surface of the lagoon. There was something very sinister about this weather, but it was wind that could shorten the take-off run of the air-craft and carry us into the southwest, adding knots to our groundspeed in the first stages of the critical race for range to Bora Bora.

We rowed ashore for breakfast where the crew of 603 were already gathered at the camp. Within a few minutes the little fresh puffs on the lagoon had changed to squalls throwing themselves on the water and spreading out in dark claws of wind that rushed wildly over the surface, hitting the air-craft and making her sway with the lift on the wing. She was swinging from northeast now, the wind changing direction as its force increased.

There was real menace in this weather. I went for the din-ghy, to go out and lay the second anchor. As I reached the dinghy and was dropping the oars into the oarlocks I saw 603 begin to move. Struck by a vicious squall of wind she swung, very deliberately, and started to drag her anchor. Then she picked it right out and walked away for the shore and the jagged rocks.

She was gone, unless the anchor snagged up on the bottom before she reached the shore.

I rowed back and lent the wooden dinghy to Spinks, 603's captain, whose plight was immediate. He got into the dinghy and laid in with the paddles, but he could not reach his air-craft in the face of the wind which now was blowing—about fifty knots from northeast and dead onshore.

I stood, my feet fixed to the rock, fascinated by the inevi-tability of disaster, unable to do anything to avert it.

And then the aircraft, drifting for a moment half across the wind as she began to bear away, suddenly swung up, and stopped. The line from her bollard stretched tight like a rod straight to the water ahead, as though taken by some enor-mous fighting fish—and held fast where the anchor had caught on the coral.

For the moment she was safe, held on the one taut line. And *Frigate Bird* was riding securely at the mooring. There was nothing more I could do about her.

With the wind came rain in blinding sheets, sweeping into the camp, drenching everything. I made a dive for the cook-house shelter, caught it as it was uprooting to blow away, swung it to face the wind and held it while Jock tried to adjust the shaking structure to some measure of security. The ferocity of the squall gave some indication that it might be a passing blow with a change of weather, so we took what shelter we could and held to the less secure parts of the camp, all of which were threatening to blow away. But soon we saw there was more in it than a vigorous change of weather.

The squall passed, but the wind hardened into a steady blow that even now was beginning to break things. It sizzled over the water and rushed through the camp. All idea of personal comfort soon passed—a relief really, because of the futility of trying to keep dry. But we held on to the fire, reducing the cookhouse shelter to proportions which appeared to have some chance of survival. The camp itself began to disintegrate round us as the wind, instead of passing, rose steadily in violence.

Within an hour of our coming ashore for breakfast, little of the camp was standing. Those of us not engaged in trying to keep the fire going crouched behind the remnants of the camp, seeking some shelter from the wind and rain.

The birds had left the foreshores, except for a few which now could not escape. Where they had gone I do not know, but I suspect it was to the Rock, where there was shelter from any weather in the caverns and fissures.

A violent gust took the remnants of the cookhouse. I was bending down now, stirring the fire to burn the wet wood I was putting on it, when the whole shelter rose off the ground and took off as I ducked to avoid being taken with it. It lifted into the air and was swept away, to fall and smash itself to pieces beyond the camp. It was followed by the fire, now exposed to the force of the wind, in a shower of flame and sparks like a comet, till in a few seconds it was extinguished by the rain.

I got up and made my way to what shelter I could find. I noticed that some of those who had arrived recently at the island were already blue and shivering with cold. Norman Birks looked like a wet eagle considering his next move for prey. I knew what he was thinking and what the others were thinking as we just stood there miserably and watched *Frigate Bird* snatching at her mooring as each withering squall struck

her and the short steep seas smacked in under her nose. But
we felt a strong sense of security from the mooring—the
enormous weight that had gone crashing to the bottom of the
lagoon—if the lines would hold. The life of *Frigate Bird* depended on these strands of hemp. I had done everything possible to protect them from chafing when we secured the gear
to the mooring. There was certainly no more I could do
about it now.

Suddenly the wind eased; dropped to about thirty knots.
Warrant officer Hicks, the 603 engineer, darted out from
shelter, jumped into the dinghy, and rowed frantically but
effectively for the aircraft.

She was about fifty yards from the shore, still holding precariously on one anchor.

It couldn't have been more than two or three minutes from
the beginning of the lull in the wind till he reached her side,
and, fending off the bouncing dinghy to keep her nose from
punching a hole in the hull, he clambered aboard and pushed
her away, letting her drift back, blowing in for the shore. Before she had reached it the wind swung into east and shrieked
down on the island more fiercely than before.

Hicks disappeared into the airplane, and closed the blister.
We saw no more of him as she hung there, shrouded in rain
and driving spray.

Pushing out from the shelter and staggering erratically
against the increasing wind, we ran to catch the dinghy before she could smash herself on the rocks. I felt myself struggling ridiculously against the hail of tiny bullets hitting my
face as I forced my way through some chaos that was neither
earth nor air nor water.

We hauled her out, clear of the bank, on which the mounting seas in the lagoon were now rising and piling up great
rolls of weed like green waves dumping heavily on the shore.
Making for shelter again I saw a young gannet, wet and bedraggled, but alive, in the crevice of a rock. He was in a bad
situation; the only bird left on the rocks by the camp. I found
a dry spot behind a rock, under the shelter of a ledge, and
put him there, without much hope for his survival. I reached
the shelter again, though the last of the camp was now gone.

I began to think beyond the aircraft, to the question of survival; because the wind was still increasing and beginning to lift things off the ground and blow them away. I tried to look out from behind the remains of the shelter. The wind took my breath and seemed to drive it into my chest. It was no longer possible to stand and walk against the roaring stream of air and water.

About half an hour after Hicks reached 603, there was another slight easing of the wind, during which he let go a line with a float on it. The float came within a few yards of the rocks and lay there, with the line apparently caught up on the coral. Alan Murray went into the water, trying to reach the float and then the aircraft by hauling in on the line, but he was very nearly drowned. Weeds closed around him in the broken water, and he had to struggle back to the shore, reaching it in an exhausted condition. In a few minutes the wind was down on us again, swinging more towards south.

There was no doubt now about the nature of this turmoil: a hurricane was passing over the island and, because of the rapid shift of wind, we could not be far from its center.

My attention was caught by a new sound of something rushing, roaring towards us, down the lagoon. The Rock was blotted out and the lagoon was sweeping towards us in a wall of water like spray from many hoses, reaching from shore to shore and leaving no line between sky and lagoon. It was a fantastic sight, as though the ocean was folding up like a carpet, and rolling down in some cataclysm of water torn and driven in blinding showers by the wind.

"Mountainous seas sometimes sweep right over the island." I remembered the words of the old *Sailing Directions* on Clipperton Island. They were not reassuring as I could hear this thing coming down upon us above the roar of the ocean. I watched, fascinated. Thundering down on us was a solid wall of glistening water in which the rain could not be distinguished from the driven surface of the sea.

Then it struck.

There was nothing to do but crouch there and try to breathe. Suddenly everything had become very simple. There was just the question of breathing; a single purpose in some strange but chaotic dream.

The worst of it passed quickly, and I looked out instinctively for the aircraft. *Frigate Bird* was lying in a white shroud of water to which there was no surface. I saw her lifting and swaying like a ghost aircraft flying through turbulent air—and like a ghost she stayed there, flying, but neither advancing nor fading into the background of her shroud. That she could still be there struck me as incredible, and my attention was so centered on this fact—that we still had an aircraft which I had already dismissed as lost—that I could not think for the moment in terms of anything else; till my eyes themselves, now instruments operating on their own account without direction, flashed me a message from 603, and I saw her going back into the mist of spray and rain, going as I watched her, quickly and inevitably for the rocks.

What the others were doing I cannot remember. I didn't even know they were there. All I saw was the white ghost of *Frigate Bird* flying but never moving, and the fading form of 603, sinking back into the mists.

I didn't think of the airplane, but of the man who was in her. Though this had been the calm water of a lagoon, it was now a raging inferno in which nothing could survive against the rocky shore. She was gone now, and so was he. The anchors weren't even checking her. She was blowing wildly for the rocks.

I staggered out from the shelter. Others must have had the same impulse, for I suddenly realized that several of us were clutching our way towards the shore where she must hit, and where Hicks might possibly be rescued. We were stung to blind, hostile action by the numb realization that Hicks, whose resolute initiative had sent him to save his aircraft, was going to lose his life.

Blown down by the wind, and staggering like drunks along the edge of the lagoon, clutching at anything that offered a hold, we had no chance of reaching the spot where she was heading to strike. I saw her at the end, almost shrouded in spray and rain, as she lifted on the last wave to crash in among the rocks. It was a dreadful sight, like seeing one of your own aircraft about to be shot down by an enemy fighter.

I saw her come in on the last wave, waited tensely for the crash.

But it did not come.

Dimly through the mist, I saw first one propeller swing over and spin, sweeping the air with a circle of spray; and then the other. Above the shrieking wind and the roar of driven water I heard the ruffling, eager sound of the engines as they sprang into life to save the airplane.

As though some invisible hand had reached down to rescue the aircraft, she stopped and then began to move away from the shore. I watched, spellbound and with intense feeling, as Hicks took her, blasting a way through the chaos, out from the shore—out, for water where he could hold her with the engines, facing up to the wind.

I could see that he was working at the controls, obviously pushing the wheel hard forward as squalls of incredible violence struck her, threatening to lift her out of the water and throw the twelve tons of aircraft like a moth against the land. A Catalina will fly at seventy knots. These squalls must have been more than a hundred. Only the suction on the hull and the tail raised high by the elevator to stop the positive lift on the wing were keeping her in the water.

I expected *Frigate Bird* to take off. She was held like a kite on the mooring, controls locked, and floating in an attitude of possible lift for the wing. I looked across at her, and saw her nose piled high with weed, on which the seas were breaking and driving back over the whole airplane. That was saving her. The weight of sodden weed was holding down her nose, stopping the seas from driving in under her bows and lifting her, saving the lines from the continuous jerks.

Hicks had 603 under control. He was not a pilot, but was using his head. Every time she wanted to come out of the water he anticipated her and, ramming the control column hard forward, let the wind blow under the elevator and lift the tail, breaking the lift of the wing. In this way he kept her in the water and held her steadily with the throttles, using whatever power was needed to oppose the drive of the wind. While the engines kept running and he did the right thing with the controls, he had her.

A wave of wild exhilaration swept over me. I leapt up and dashed my way to the edge of the lagoon, shouted to him at the top of my voice and tried to wave him encouragement. The wind took my voice and blew it back in my face. It

struck my body and flattened me to the ground. I clung there
to the rock, trying to shield my eyes so that I could watch
him with 603.

I saw in my mind his action in the aircraft. Most likely he
had set up everything for starting when he reached the airplane; set the throttles, propellers, and switches in the captain's station, and his own mixture and engine controls in the
engineer's station, in the tower under the wing. Then he had
waited.

When he saw her pick up the anchors and go, he had gone
into action, disregarding altogether the natural urge to save
himself by going back to the blister compartment and trying
to jump ashore when she hit the rocks. As she drifted he had
energized the starters, letting them wind up till he had judged
it right to engage them. Then, each engine in quick succession. They had fired and started.

The first thrust from the propellers had checked her momentarily while he leapt down from the tower and forward to
the pilot's cabin. He had grasped the throttles and given her
the power that had drawn her away clear from the menace of
immediate destruction. Then he had flown her on the water,
working the air controls to hold her down and keep her head
into wind, and the throttles for the power needed to draw her
away and hold a position out from the shore.

It was precision work—the one absolutely right action.
Hicks had been there to save his aircraft, without regard for
his own safety. Crouching there in the maelstrom of these
unreal surroundings, I resolved that he would receive recognition for his action.

The wind was still blowing with hurricane force, but both
aircraft were surviving: *Frigate Bird* a ghost hovering
strangely in the mist of rain and spray that blew off the surface of the lagoon, the whole of her bow section now piled
high with streaming weed; 603 well out from the shore, holding on her engines, with the slather of her propellers reaching
our ears like some faint and distant sound above the roar of
the hurricane. It was a noble sight—the two aircraft standing
unshaken in the face of the terrific forces that were driving
endlessly and relentlessly upon them. There was something
uncanny in the vision of *Frigate Bird*, the still blades of her

propellers standing starkly against the wind, holding there
with no visible means, the wing just hovering over the maelstrom of weed and water that had been the lagoon, unmoved
except for one of her engine covers, now ripped open and
flaying the side of the cowling.

The day was now approaching noon, though there was
nothing to show it but the hands of my watch, still going on
my wrist. The wind, moving round the cycle of the storm,
had swung into south and was now blowing over the narrow
strip of low land that lay between us and the sea. Though it
showed no sign of abating, the drive of the seas on the lagoon was less ferocious. The situation so far as the aircraft
were concerned was thus improving. But our own was deteriorating.

Over by the coconut grove on the very low land the heads
of the palms were streaming back in windblown fronds like
the hair of a girl in the wind. The taller palms were bent like
tightstrung bows, and held till it seemed that their spines
must break. Their fronds were awash in the blast of air,
shrieking and calling defiance to the wind and the black menace of the cloud.

And now, instead of the dark horizon with cloud against
the flat rim of the land, there was a cold white stream of
breaking surf, visible above the land and roaring thunder-
ously at the island.

The wind had worked its way into south only within the
last half hour, making of our side an increasingly leeward
shore to the ocean. If the sea had run so high in so short a
time, where would it be tonight? Already only the reef was
stopping the rollers, which I could see were higher than the
land.

If the hurricane wind worked into the southwest and continued to blow into the night to the dead lee shore at our
strip of the island, I felt fairly convinced it would come over
the land where we were now located. The only secure place
would be the Rock, which we could not reach because of the
violence of the wind. To go fifty yards was a struggle. Two
miles would be impossible.

A place which promised some measure of security was the
coconut grove, only about two hundred yards away. The

palms had certainly stood for many years, and must have weathered severe storms, though possibly none as violent as this. I decided to try to reach the grove, look over the position there.

To go alone seemed to be bad tactics. The terrific forces attacking the island suggested a potential situation in which one of us alone might be injured and unable to return. I asked Bligh to come with me, shrieking at him: "Going to have a stab at reaching the palm grove, Len. Like to come?"

His reply was snapped up by the wind, but I saw that he wanted to come.

We first made for the ocean side where, over the bank, I thought we might find a cushioning effect upon the wind. We reached the pebbled slope, but found the sea already washing so far up the beach that to go near it would have been too dangerous. We struggled and crawled along the top of the bank, holding fast to rocks and gaining ground as we could.

In half an hour we reached the shelter of the huge iron mooring buoy that lay near the grove about two hundred yards from our starting point. But we could not go beyond it. We had barely reached this refuge when the weather came at the island with another attack of such primeval savagery that we had no more thoughts of the palm grove but were glad to shelter behind the buoy; feeling that in a world which now must surely disintegrate, we for the moment had a place of temporary refuge from which we would have a good view of the performance.

My first act behind the buoy was to pile up some large stones, jammed well in under its round belly, to stop it rolling over and squashing us like a steamroller. The whole situation was so fantastic that it amused us. It was now far beyond the stage where there was anything worth worrying about or viewing with any seriousness at all. We were down to fundamentals, where it was wonderful to be able to stand up without being blown down immediately by the wind.

With this last blast of the hurricane I had dismissed the aircraft. We could barely see the few yards to the camp, and out where the flying boats had been riding was nothing but a sheet of white spray and vapor through which we could see nothing. I was completely resigned to the fate which I was convinced had now overtaken them, and Hicks.

The aircraft, the flight, our whole purpose of being at the island, now belonged to the past and were already put away. All that concerned me was the present, and the future only because I was aware of the fact that, somewhat inconveniently, it would become the present and therefore would have to be provided for. To put a stone under the rusty buoy to stop it rolling upon us and killing us was of exclusive importance. When that was done, looking to the coconut grove and to the sea was the future.

Bligh and I laughed at each other and at the absurdity of our situation. We felt deliciously free now that we were relieved of speculation, but the menace of the rising sea began again to impress itself upon us.

It was roaring past now not two feet below the top of the bank. The surge was sending little rivers trickling through the stones over the top towards us. Now there were no individual rollers breaking on the reef. The whole ocean was tearing by in a roaring flood of water, clawing at the island. The palm grove, which had appeared to be a place of security from down by the lagoon, had no more real entity than straws in a flood.

Wild pigs were there, and I could see some birds crouching half stunned against the ground behind the stems of long-fallen palms, under pieces of rock, in bunches of tangled growth—anywhere they could find shelter from the storm. I thought there was some significance in their coming here rather than to the Rock. I again began to build up some confidence in this grove as a refuge for the night, and began to plan how we might all reach it, taking with us as many of the remaining stores as we could carry.

My plans were changed by the weather. About three o'clock in the afternoon there was a definite lightening in the sky, and, I thought, a slight easing of the wind. It had the effect of increasing the menace of the sea, the sinister character of which now was the dominant factor of the elemental forces attacking the island, where before, the ocean itself was blasted to submission by the wind and rain.

Then, suddenly, we caught sight of the aircraft. Both were

still on the lagoon, 603 facing up with her engines still running; 275 lying, inevitably it seemed, on the mooring. We
watched them with intense wonder, both of us, I think, re-
constructing an existence we thought had passed.

In a few minutes we saw that the wind really was slacking
up. It was still blowing what we would normally have
thought to be a screaming gale, but the unbelievable violence
had gone. We decided to make our way back to camp. I saw
a possible chance of reaching *Frigate Bird* from there.

When we reached the shore of the lagoon, Birks was in the
water. Having a slight improvement in the weather he had
tried to swim to the aircraft, where the starboard engine nacelle was being belted unmercifully by the torn engine cover
and was likely to be damaged or to have something pulled
away.

We afterwards learned he had managed to struggle out
nearly to the aircraft, swimming in driving seas loaded with
masses of weed. Then, realizing he could not make it, he had
turned back for the shore. Bligh has reached him just as he
struggled to the rocks and, completely exhausted, he lay there
for some time trying to recover his breath. Tough as he was,
it was some time before Birks fully recovered from this experience.

I could see that to secure the engine cover was now important and urgent. The cover was heavy canvas with metal
rings, and it was obvious that something would go if it continued to batter the cowling. Also, it was causing considerable
resistance to the wind and extra drag on the mooring lines,
about which I didn't want to think too much.

Near the site of the camp was a small promontory of rock,
which now was almost exactly upwind from the aircraft. While
there was obviously no possibility of rowing the dinghy out
across wind, nor, of course, against the continuing wind of
more than sixty knots, I thought it might be possible with the
wooden dinghy to get down to the aircraft from this promontory. She might be maneuvered to pass close by the hull, and,
as she was passing, one of us could grab the wing bracing
strut and slip a line around it quickly, thus securing her for
the moment. Then we could climb out onto the strut and let
the dinghy go on a line so that she could not bump the boat,

and moor her there; then climb up on the aircraft and go to work on the engine cover. That seemed reasonable.

Four of us carried the dinghy to this point and we prepared it for the attempt to reach the aircraft. In case the oars would not hold her, I tied one of the small drogues from the rubber dinghy to a line on the nose, took in the oars, and sat in the middle seat ready to use them, while the others held the dinghy bow on to the seas.

Leaving Birks to attempt a passage when possible with the others in the rubber dinghy, Henderson and I set out for the aircraft. I soon found that the drogue was not necessary. With Henderson's weight in the stern, and the bow thus well out of the water, most of the seas ran by her and she was quite controllable with the oars. We let her blow back, stern first, for the aircraft, checking her to stop her filling over the stern.

Facing the aircraft as the dinghy was blown back towards her, I tried to make an accurate approach. If we lost the position dead upwind we'd never recover it. I took the weight on the oars, checking her a bit this way, then that, aiming to let her blow by the aircraft where we could reach up and grab the strut. If I missed it, and Hendy missed it, we'd "had it" as far as reaching the aircraft was concerned. We'd just blow on and pile up on the rocks of the shore.

To be doing something was a relief after hours of submission to the elements. Just to feel the pressure of the water on the paddles, a sense of control coming back through my arms, and the firmness of my feet braced against the tuck of the dinghy, changed the character of the storm. We swept down under the wing and managed to grab the strut as the dinghy was driven by the hull.

There we hung on, and somehow climbed aboard, letting the dinghy go away on the line astern. The rank, raw smell and driving wetness of the hurricane were on the aircraft.

Hendy and I climbed up the side and into the blister. The hull was dry inside. I was conscious of our bedraggled condition: here in the cabin where a dry suitcase with some hotel labels on it lay on a bunk. That such things could exist seemed unreal. We went forward and out through the roof hatch, again blasted by the wind.

In a few minutes we had the engine cover lashed down and everything snug. As a last precaution I put out the second anchor. It wouldn't have held at all in the ordinary way, but there was just the chance that if the mooring lines went, the anchors might snag in the coral and save her now that the force of the wind had decreased to that of a gale. Both engines were out of action, the port with a defective starter mesh and the starboard with some of the connections not yet made. The wind was still too strong for us to attempt any work on them.

We went below, put on some dry clothes, and got some coffee going on the stove. I set the altimeter back to zero and kept a check on the pressure movement, which was soon shown to be rising steadily.

The progressive shift of wind from north in the morning, through east, to south was normal, and also indicated the passage of the storm center. For the first time I really began to allow myself the luxury of believing the aircraft was going to survive, but this was held by the threads of the mooring lines, the condition of which we could not know.

By five o'clock in the afternoon the wind had dropped to a hard blow of about forty knots. Hicks still held 603 on the engines. I could see Birks, Bligh, and Hogg floating off the rubber dinghy into position to try a downwind passage to the aircraft. The dinghy, tied down to some vine roots, miraculously had survived the hurricane.

They let go the shore and, paddling furiously, managed to keep her drifting to pass by the starboard strut. In a few moments they were there, blown skidding on the waves, and passed us a line which secured them to the aircraft.

Soon afterwards the crew of 603 reached their aircraft, but continued to hold her on the engines against the wind, which was still too strong for safety with the anchors alone.

Our situation at the island now was on the way to being retrieved. From one in which the most optimistic hope seemed to be limited to that of our own survival, we were now established again aboard the aircraft, which apparently had suffered no severe damage. Everything ashore was wrecked, but the canned provisions were mostly intact and I

felt that if the night held fine we should soon be well established again.

We cleared the weed from the bow of the aircraft and,
when we were able, hauled up to inspect the mooring lines.
The heavy rope had been stretched to thin, tight lines, hard
as metal rods, by the terrific stresses from the driven aircraft,
but all were still fast to the mooring and, as far as we could
see, undamaged.

As night came over, and the wind eased to about thirty
knots, 603 went onto the anchors and stopped her engines.
Hicks had held her there for seven hours, during five at least
of which he could not have relaxed for a moment. His had
been an inspiring performance.

We gathered to relax and stoke up with hot food and
coffee in the comfort and shelter of the lighted hull. It had
been a terrific day and we made no plans, simply enjoying
being together again with the airplane.

Afterwards, out in my bunk, there was peace. I listened for
a while to the dying wind, sensitive to the possibility of any
increasing sound; but, with the island, I was soon asleep; and
knew no more till I saw the sun shining in through the perspex in the morning.

Chapter 14

The Race for Range—and Bora Bora

An occasional crystal shower drifted over the island, and the breeze was light, still from south. The lagoon itself was calm and tranquil; and only the low thunder of the ocean remained to tell of yesterday's chaos. We took off the engine covers, and when the showers had passed we laid out everything to dry in the sun.

That the aircraft was undamaged was not surprising; because she had merely experienced on the water conditions through which she had passed often in the air: a stream of flying water in a wind of more than a hundred knots. She had been flown like a kite, held by the lines to the mooring, the mass of weed on the nose keeping her down, and the wingtip floats instead of the ailerons checking her lateral movement.

Ashore on the edge of the lagoon we found the young gannet which I had rescued during the hurricane. He was very much alive. The whole bird colony was a shambles, but the gannets and terns appeared to be unconcerned and were fussing and preening themselves in the sun.

171

Our own camp had been demolished, entirely swept away and scattered in the sea.

By evening we were able to run the engines. They started easily and we let them warm up and relax with the aircraft on the mooring.

I did not take her out into the lagoon for a run-up test this evening. That would have been stretching our luck. Until the engines were proved, the mooring represented our only security against wind. I wanted to see a fair day ahead before subjecting the aircraft to any risk of an engine failure. There was no single reason to expect trouble with the engines, but the accumulation of circumstances through which they had passed since we had last used them added up to a situation in which I felt they could not be relied upon until a vigorous run-up on the lagoon had chased the gremlins out of the cowlings. I hung firmly on to the mooring till we could see a fine day, with light wind ahead of us.

That night Bligh handed me a message from the radio. There was a smile on his face as he watched for my reaction.

Two Dakotas now loading up with engineering equipment at Montreal were coming to the island on an early date mentioned, with a party of engineers to start work on the base. (We had found on the island a strip of level land where the phosphates had been worked, leaving a natural runway long enough for this type of aircraft.)

Our reaction to this encouraging news was briefly recorded in my log for this day: "There is great satisfaction in the knowledge that our work is to be followed up by this early development of a base at Clipperton Island."

Secretly I had dreaded the messages coming over the radio each night. Though all the communications with 45 Group Headquarters clearly indicated a sympathetic and intelligent understanding of events at the island, and I was confident that the same would be true of Transport Command Headquarters, I had little confidence in any stability beyond that point.

It so happened that my presentiment was well founded, but we escaped from Clipperton Island before the political chameleon was able to change its colors again.

On the morning of the 14th conditions were perfect. The

weather was intensely fine, with a light southerly wind. I decided to leave at noon, a convenient time which should bring us to the Marquesas Islands the following dawn, and to Bora Bora before the end of that day.

Nothing could be gained by making a separate operation of the engine test, involving a return through the coral to the mooring for filling the tanks, so I planned to make the test on the run-up before departure. So we brought up the fuel raft, topped up the tanks to the filler caps, and left our friends of 603.

As a final gesture, the port engine starter would not engage and the A.P.U. would not start. But neither of these defects was vital. Hogg spun up the inertia starter on the main batteries, Birks stood up on the wing and engaged the starter by hand and the engine fired. With some agility he slipped back over the wing and in through the blister instead of being blown off into the water.

The starboard engine started immediately and I gave the signal to Henderson to cast off the mooring. He let go, leaving the ends of the line afloat on the buoy for 603. I gave her a burst of starboard engine to swing her away from the shore and we moved away, picking a track out to the clear water. It was fine to have the aircraft alive again, to feel her shoving through the water, and to have the power in my hand with the throttles. I took her out over the edge of the crater into the deep, and let her go round, circling in safe water, to warm up the engines. She was low in the lagoon and heavy, pushing a rising bow wave ahead of her.

On the word from Jock I gave her the power, checked the propellers, then ran up the engines to thirty inches. She held it well on the port engine, on each magneto. Then I tried the starboard. When I switched off one magneto the engine vibrated, misfiring and dropping about three hundred revolutions. I tried her on the other magneto alone and she held the revs. Then I had to shut off, before the end of the deep water.

I took the aircraft back to the Egg Islands, let her swing upwind again, and gave her full take-off power. It didn't clear the ignition drop. The engine still misfired badly on the left magneto. The inevitability of return to the mooring stared grimly at me. I accepted this reluctantly, and headed her back for the bay by the camp.

The cause of the misfiring was not obvious, though it could have been faulty plugs. The thought of removing more plugs was a bad one and a depressing anticlimax to our high spirits as we had headed out into the lagoon to take off on the flight for Bora Bora, something for which we had now been striving for nearly six weeks.

But the fact that the starboard engine was unserviceable had to be recognized, and, in the present condition of this engine, failure of the single ignition system of the right magneto would almost certainly mean coming down in the ocean, unless it happened when she was light, towards the end of the flight.

We picked up the mooring, cut the motors, and Jock quickly located three cold plugs on the left bank of the starboard engine. He changed these, and with less than half an hour's delay we started up the engines again and took her out to the open water.

I ran up the motors and turned over onto the left magneto of the starboard. She was still misfiring, as badly as before, dropping several hundred revolutions. I quickly turned her onto both magnetos, and gave her full power in another attempt to clear the engine. Then I shut her down and taxied back to put into effect a decision I had made when we left the mooring the second time.

This was the one time that I did not invite comment before making a drastic decision affecting us all personally. I purposely deceived the others, except Birks, who could see what was happening, anyhow. I turned her and ran up the engines again, this time taking care not to run the starboard on the left magneto alone.

I called through to the engineer, "All O.K., Jock?"

"Aye, Skipper. She's fine. Everything normal now."

"Right, Jock, we'll get away."

This decision screamed at me in outrage of my own principle against the acceptance of avoidable risk in the air. I was convinced that if we didn't leave now we would never leave. For weeks we had struggled with an assortment of adverse influences, and had somehow managed to keep afloat. But I now had a strong impression that the aircraft was gradually bogging down in the insidious effects of wind and weather and something that inhabited the island, and that failure to extricate her now, in what favorable circumstances we could command, would be the end of it.

This impression was supported by the facts which were now before us. It seemed that changing plugs was not going to smooth out the starboard engine.

Sometimes there seems to come a stage when the only way to bring life back to an aircraft is to fly it—and fly it hard and far. And sometimes this is true, for such a reason as saturated ignition leads, which may dry out in flight, and this was a possibility in which I now had some faith.

As we came to the end of the run-up, Norman and I looked across the cockpit, obviously with the same idea. I leaned over and called, "We'll test it again at Bora Bora."

To tell the others that she was still missing on the one magneto would have been superfluous. I knew they would agree to go, and they might as well have the comfort of thinking that the last run-up was genuine.

I let her rumble down to the red flag mark we had laid in the corner north of the Egg Islands, and there brought her up to wind. Ahead was a run of a mile and a quarter before the Great Reef. Then there was the over-run, on a turn, for the small deep by the Rock. By heading for the coconut grove, and going into a turn on the water, the length of run could be increased. Birks was to handle the power, and we had agreed to run her up to fifty-two inches, four inches above the maximum permissible forty-eight for take-off, if she looked like needing it, which we knew she would, in the very light breeze which was blowing.

We cleared her for take-off and I gave Birks the word for the throttles. "All right, Norman; give her the power." Then I forgot about the throttles and went with the aircraft. It was wonderful to feel her tear into the take-off.

At first Birks gave her only enough to move her away, to start her plowing heavily through the water. Then, in a few yards, she began to smash her way to freedom, blasting the island with thunderous sound and sending a deluge of water

over the nose so that I could see nothing beyond it, and had
to go onto the gyro for direction.

Fed with the power as she could use it, the aircraft finally
reared up on the bow wave and I began to see ahead, holding
her for the grove. Soon I could feel her wanting to ride over
the wave and go. I snapped the floats' switch onto the signal
for Jock, and as they retracted to the wingtips she put down
her head and went.

I felt air control come through the ailerons, and as she
picked up speed I eased her gently into a turn, taking the
weight on the aileron to hold the wingtip just clear of the
water. She was going all right. I knew it then, by the clear
run of the water and the tightness of the airplane. As I
straightened her up for the run down the western reefs, I saw
Birks' hand go forward with the throttles, giving her everything, now that she could use it.

She blew the water under the hull, riding clean, but not yet
flying. I could see the stain of shallow water on the Grand
Récif coming closer, rushing in towards her now; but she had
to go. She was confident, singing with a million voices in
high-pitched harmony to take her away over the reef. I
reached up and gave her a touch of tail trim, felt the fine
balance of approaching flight. The wing had her now, knowing it could take her from the water.

I made no effort to drag her into the air, but rode with her,
letting her know I was there and ready to help her break
away when she was ready. As the shallows of the reef swept
in below, she had the water beaten. It was time to go. I took
a light tension on the control column and lifted her away.

She was heavy in the air, but flying securely, as I held her
down for speed. Brushing the last of the water from her hull,
the air took her with smooth release to a sudden freedom
from the sea. I felt the current of that freedom flow in to
center on the aircraft and, with some intense high frequency,
charge me through the touch of my hands on the control
wheel. I called for Birks to reduce power as we swept low
over the sea, and leaned her into a turn to come back and fly
over the camp where the others were standing, watching the
take-off.

I poured her down over the lagoon, and, in a steady turn

that she would take with the heavy load, saw them waving up to us.

All that had happened at Clipperton seemed suddenly to become condensed and unimportant and, as I looked down for the last time upon the island, to sink with its strange past into the blue depths of the crater. Now there was no reality but the airplane: Norman Birks there in the starboard seat, Hendy leaning over the chart table, his chronometer watch resting in the nest of his upturned cap, his gadgets arranged ready for the job before him; Len Bligh with the earphones on, tuning his radio to start his long vigil in a world of sounds; Jock, I knew, watching closely the engine instruments in this early stage of flight.

I glanced up to the motors, saw them filled with strong significance of power, leading the wing that followed with smooth contentment, but floating heavily on the air and serious with the early responsibility of its great load. I continued to fly her by hand till we found the level we wanted.

With three thousand nautical miles of virtually unknown winds and weather ahead of us, the flight for Bora Bora was a critical race for range to reach this island.

Now, in the beginning the surface wind was dead ahead and to be avoided immediately if possible.

While we were preparing to leave I had lain on my back on the wing, watching the cloud movement. The few wisps of low cumulus were moving with the surface wind from south. Above these there was more definite cloud of the same variety, fine weather stuff with tops at about four to five thousand, sloping away from north and appearing to be swept that way by a reversal of wind. I decided to go at once for this wind, and drove her hard, with plenty of power, till we reached it.

For every other reason it would have paid to let her loaf along at a thousand feet, saving the engines till the starboard had had a chance to settle in after being partly dismantled, and burning only the fuel of cruising power till she was lighter and more willing to deal with height. But I could not afford to waste any opportunity to reach a favorable wind, nor to avoid an unfavorable one. Against my own sympathetic reactions for the engines I had to push her for the height.

She broke out through the top of the cloud layer at four thousand feet. I let her go on up to four-five to cover the scattered tops, then shut down the engines to cruising power.

At two thousand r.p.m. and twenty-nine inches of manifold pressure she held the height in a good flying attitude, and this was the lowest power combination at which this was possible. Any heavily overloaded Catalina usually needed more power than this in the early stages for maximum range, but 275 had been ruthlessly stripped of everything that could be taken off to smooth out the airflow and to reduce her basic weight, and she was going well with 595 horsepower from each engine.

Right from the start this flight was to be a keenly contested game with the elements and the aircraft.

Like other Catalina captains faced with the long flights on narrow fuel margins, I had discovered and recorded the speeds at which, in practice, the aircraft should be flown according to her weight as she burned down the fuel. Experience taught us the power needed for these speeds, and this formed the basis on which I now set the power to see how she would take it, and found that at two thousand r.p.m. and twenty-nine inches of manifold pressure she was riding on the air, showing 105 knots on the airspeed indicator. This was the lowest speed at which I could expect her to fly without sagging along squelching through the air and doing no good.

For working ease I had adapted to the engine power curves combining r.p.m., manifold pressure and altitude, a scaled instrument on which I could read off immediately the power resulting from any combination, with allowance for departure from standard atmosphere, on which the power curves are based. With the power read off in this way I could enter a graph of fuel consumption against horsepower, which I had made up from previous observations of fuel actually consumed in flight.

Both were simple arrangements, accurate and easily read, and therefore more feasible for use than elaborate volumes with columns of figures, graphs, and different colored lettering, covering innumerable theoretical circumstances and, because of their complexity, often full of dangerous mistakes.

To extend the range as far as possible westward we went after air that was moving that way with us; and to find it was essential if we were to have any margin at the other end. The means of finding it from signs in the sky, on the sea, and from our navigation would develop as we flew.

Over this middle cloud I could see the tops sloping from northeast, and down on the sea the white-capped surface driving up from south, streaked by a southerly wind the force of which I estimated to be about twenty knots. So, even with as little height as 4500 feet we were in more favorable air flow. But whether the wind was with us from northeast, or was from southwest but less up here than below the cloud base, was not visible; because cloud tops sloping against the lower wind may indicate merely a lag of the tops caused by wind from the same direction but of lesser force than that below.

There was the evidence at the island which had showed the higher reversal to northeast as I had watched the cloud movement, and there was the slope of the cloud top, all of which was favorable. Rather than take multiple drift observations to find the wind force and direction, I decided to hold on at this level and see where we fetched up later, when the sun would give us a line for distance run. The deviations from course of even double drift observations were to be avoided, especially early in the flight. It is possible, however, to find the force and direction of the wind in which an aircraft is flying by observing with a drift meter the angle of drift on two or more widely separated headings. This can be done by sighting on the white patches of wave-top foam, which are relatively stationary on the water. The drift recorded from these observations, applied with the compass headings and speed of the aircraft through the air shown on the airspeed indicator, can be resolved to give the wind force and direction at the aircraft's flight level. She was settling down well, using slightly less than six hundred horsepower per engine, and burning, according to my consumption figures, eighty-five U.S. gallons an hour, making a true airspeed of 114 knots. At this rate of consumption we could not reach our destination without an appreciable average following wind, since it would give slightly less than twenty-five hours endurance in which to fly three thousand nautical miles. There was little significance in this, however, because the consumption could be brought down considerably by reducing power with the reduction in load as she consumed the fuel. According to my reckoning and experience, it could be reduced to little more than half this consumption at her lowest weight, when the boat would whistle along on a smell of fuel, flying well at a hundred knots.

Allowing for the possibility of having to use high power in rich mixture to deal with weather for some of the flight, I reckoned that we had a safe thirty-hour range, which was a better picture. It was, nevertheless, one in which every point had to be scored to cover unusual developments, such as the loss of an engine when, though she might be down to a weight which could be supported by single-engined performance, the consumption would be high for the distance gained at the reduced speed.

I felt that things were going so well it was almost worth trying the starboard engine on the port magneto to see what would happen; but I resisted the temptation to do this, because I might find her still misfiring. After a few moments of speculation, I decided to test it, as we had agreed, before coming in to land at Bora Bora. I handed over to Birks and went below to make contact with the other functions in the aircraft.

We were on a course for a large rock on the northeast fringe of the Marquesas Islands, 2100 miles from Clipperton. Henderson, as navigator, had a formidable responsibility on this flight, though his problem was simpler than that of finding Clipperton Island. But the result of error would be more serious.

To reach our destination and, in fact, to reach land at all, he had to be exactly right in the work that was ahead of him. He could not rely upon radio aid, and we could not return to the broad front of a continent if, in the later stages, we encountered conditions impossible for accurate navigation. There was radio at Bora Bora, but few aircraft had used it, and recently none at all, according to reports.

I knew that Henderson had the ability and experience to make the landfall at this rock, two thousand miles away, and subsequently the island of Bora Bora, nearly another thousand, but that flight conditions could make the successful application of that ability and experience very difficult and, in some circumstances, impossible. To give Hendy the best conditions, we had to apply a sympathetic appreciation of the work that was ahead of him to our handling of the aircraft.

It is a surprising and dismal fact that the most persistent influence with which an air navigator has to deal is steering error. When he has made his allowances for variation of the compass due to earth's magnetism, for deviation due to its effect through the iron in the aircraft, and for the drift of the air in which the aircraft is flying, he still has to contend with the fact that the pilot may not steer the course given to him. In fact, it is rare for the course to be steered accurately, and still more rare when the automatic pilot takes over the steering. Precession of the gyro, on which the auto-pilot depends for its control, causes a tendency for the aircraft to wander in one direction, and this has to be checked regularly by the pilot on watch, and corrections made by adjusting the auto-pilot controls. To keep an accurate course, this often has to be done every few minutes, depending mainly upon the course of the aircraft in relation to earth's axis of rotation—and, of course, on the condition of the instrument.

The reliability of the auto-pilot itself, and the fact that it has relieved the pilot of the long hours of concentration when flying the aircraft, naturally cause him to relax, and unless there is a real appreciation of the need for accuracy in steering, the aircraft is allowed to wander. This is most likely on long ocean flights where a pilot is often on watch for many hours, and he must relax to retain the freshness necessary for the final approach to his destination.

Working the aircraft is a compromise, in which the navigator is usually the victim. Generally, the pilot is fairly consistent in his steering error, though he seldom realizes it himself. A wise navigator, who wishes to avoid the intolerable exasperation of frequent steering corrections, watches his master compass, soon notes the average steering error, and accepts it as another allowance which has to be made for the compass course.

This procedure is, of course, wrong in principle, but it

is a practical and often necessary compromise. The pilot on watch is usually occupied, and his side of the compromise is to strike a reasonable balance between continuous concentration on the instruments and everything that is happening in his aircraft and the air in which it is immersed. The pilot who has nothing to do is living in a world of fiction, which he is unlikely to inhabit for very long.

The result of these circumstances is often distressing to navigators who sometimes, because of their natures, seek perfection, and seek it relentlessly as the only means of fulfilling the urge that is in them. Sometimes such specialists cannot see the view of the pilot, who is often a man with a wider acceptance of the inevitable, perhaps acquired by sitting often for long hours at night over the ocean, knowing that the orbit and the very existence of his world depends upon the engines that look to him from the wing.

The navigator, living with stars, a chart, instruments, and books of tables, all of which are concrete things the existence of which is definite, and inhabiting the lighted cabin which also has an impression of being infallible on its own account, is often less aware of the fallibility of his world than the pilot of his. It was not difficult for me to appreciate Hendy's needs; I well remembered my own flights as navigator when conditions were almost exactly similar to those in which Henderson had to find the way to Bora Bora.

I have said that Henderson's problem on this flight was simpler than that of finding Clipperton Island. By that I mean that Bora Bora, one of the high islands, could normally be seen at a greater distance than Clipperton, and therefore was, in effect, a wider mark to fly for. Its great distance away across the ocean mattered little to a navigator of Henderson's ability. Unless we ran into phenomenal weather, extending over a long distance in the later stage of the flight, he would fix our position from star sights within a thousand miles of Bora Bora, and we would set off from that position as though we had seen and identified an island.

From that position we would fly for Bora Bora, narrowing down the last stage for the run in by position line from the sun, or by checking off on the Tuamoto Islands; or both.

Technically, therefore, the problem was an easy one for a

good navigator. The process of its solution was not easy. He had to keep alert for up to thirty hours and he couldn't afford to make mistakes in calculation, star identification, drift sights, nor in all the long train of simple arithmetic his mind would have to dictate to the pencil on his paper.

Fatigue resulting from too much work in the beginning had to be avoided. Henderson was aware of this and shared with me the principle of letting the airplane go without much interference for a thousand miles or so. All I wanted from Hendy, in the early stages of the flight, was the information necessary to convince me that we were flying in the general direction of Fatu Huku, making as good use of the wind as possible.

The effect of this wind can best be seen by imagining the aircraft immersed in an invisible ocean. Whatever our heading and speed through that ocean, we travel also with the current in which we happen to be flying. If the whole ocean were flowing in the opposite direction to that in which we were flying through it, and at the same speed, the aircraft would make no progress. Wind is a stratum of the ocean of air moving in relation to earth. It cannot move away and pile up in one place, leaving a vacuum. Air, like water on the surface ocean, is always seeking to stabilize itself, sometimes flowing back at one level to replace air which has moved away at another. So, by searching for signs of this movement, it is often possible to discover a layer of air that is flowing in the right direction, when signs on the surface indicate a head wind.

During the afternoon we were able to confirm, from observations of the sun, that flying just above the cloud top the wind was behind us, adding five knots to our speed through the air. On the water, long swells were rolling up from south, and still showing the streaks of the southerly.

About four hundred miles from Clipperton Island the cloudtop began to rise and the sky ahead was streaked with thin layers beyond which I could not see.

I climbed her to 6500 feet, where she appeared to be flying to clear the main tops, and let her go on at that level.

The engines were running perfectly and we were able now

to bring down the power to 565 brake-horsepower per engine, and thus the consumption to seventy-eight U.S. gallons an hour, while still maintaining a speed at which the aircraft was flying cleanly.

We were careful to limit our movement in the aircraft and thus to maintain the delicately established trim. Going back to take a drift sight through the tunnel hatch tended to upset the trim, and usually lost us five knots of airspeed, which had to be picked up again by letting the nose go down till she was really flying again and then gradually ease it up to level flight. Then she would hold it, till somebody had to move fore and aft again, and this procedure had to be repeated.

Another influence upon the speed, and therefore the range, was the fixed load distribution. This we had arranged so that in level flight she was slightly nose-heavy with neutral tail trim. Then, by turning the trim backward till she was balanced fore and aft, she seemed to put her head down and go. Contrarily, with tail-heavy basic loading she would have sagged along, leaning on the air.

Hogg was keeping a close watch on the engines, particularly the starboard, and all the temperatures and pressures held normal. His engines, on the fuel flow meters, were showing a consumption steadily two gallons per hour each engine less than my calculations from r.p.m., manifold pressure, altitude, and temperature, and hence from brake-horsepower to consumption, from my chart. This was on the right side since I knew my deductions to be, if anything, slightly pessimistic.

After three hours we had used 330 gallons of the 2130 with which we first left the mooring, but this included the running on the lagoon, returning to change the plugs, and the final tests on the water. So it did not represent the rate of consumption in the air. The 1800 gallons left in the tanks was all good fuel now that we had some height and were well settled down.

Radio communication also was going well. Bligh had contacted the base aircraft immediately we were in the air and was in touch on 6440 kilocycles.

The prospect before Bligh was continuous watch, if possible to maintain communication right through, to keep Dorval advised of our progress, and to be in a position to tell somebody if we happened to be forced down in the ocean. The latter was unlikely, but Bligh's part was one on which our survival might depend if we succeeded in landing intact; because, though no ships or aircraft were on our track, some action would be taken if it were known that we were in the sea, and where; and this was, at any rate, a technical safety measure.

The stratus cloud layers through which we passed before evening thinned out and we left them behind at sunset to see ahead into a clean sky with nothing visible below us but a sea of cloud stretching to the horizon. The view on which I looked from the pilot's cabin was, in itself, complete fulfillment and symbolic of our purpose. The air was new, untouched by anything, and it was receiving us with gladness.

We had passed beyond earth and no longer belonged with any world. Solitude was absolute, but inhabited by the fundamental source of life. There was no life itself, nor death, nor any passing state, but only eternity without time, distance, nor any dimensions. There was no aircraft with engines and propellers thrusting their way through the air—and no crew. I was momentarily conscious only of the source of all things of which we were part. Here in this new sky, colored with strangely beautiful lights, was the revelation which could not be named or expressed in any terms of reason, but only as a sublime consciousness which I recognized, in its fulness, as something which before had touched me with a quick impression in the solitudes of the air.

Held in suspense with this new reality, I watched it pass with the fading of the light, till again the motors were roaring above us, I could hear the click of the auto-pilot, and see the first of the stars.

I saw Birks reach across to adjust the turn control to keep her on the course, and I looked down to the lighted cabin where Bligh was taking a message from the radio.

Hendy was leaning over the chart table, smoking and penciling in the vertical section of the cloud for the weather record of the flight. I called down to tell him the stars were out, and left my seat while he came up to take some sights.

Inside, the aircraft was homely and warm. Jock switched

on the stove and we heated some food and brewed coffee. In another hour he would begin to transfer the fuel.

The position from these first star sights showed a falling off of ground speed to an average of ninety-eight knots since the last sun line late in the afternoon. That was bad, and it meant we had to go after a following wind.

Before changing altitude, double drift sights were taken to find the force and direction of the head wind for the weather records. Immediately she was back on the course from these drift observations I let her go down to search for the top of the low level wind. Now that the upper wind was from west I expected an easterly component in the lower levels.

All the signs of the night indicated that we had passed from the areas of land influence into the more stable weather of the wide open ocean, and could expect the airflow on which I had based the range of the aircraft to be sufficient for this flight. Above, the night was clear and the stars intensely bright. Below us were the gray shadows of scattered cumulus cloud typical of the easterly trade winds.

By taking a succession of double drift sights on flares dropped from the tunnel hatch, we felt our way down for a reversal in the direction of the wind.

The westerly held, decreasing slightly in force, till we descended to three thousand feet and were in the cloud layer. Then results of the next observations, taken through gaps, showed a swing of the wind to east-northeast, dead behind us. She was away again, and piling up the score that we needed.

But the air was rough, and bad for navigation and for flying. It was too turbulent for accurate sights with the bubble sextant, just picking off the stars through the open spaces between the cloud, and it upset the trim of the aircraft, making her wallow and need more power to keep her flying.

Against these disadvantages we were gaining five knots on the airspeed, instead of losing fifteen, as we were up at six thousand feet. To climb higher would certainly have run the aircraft into an even stronger head wind, because the westerly would reach up beyond not only the economical flying ceiling of the Catalina but far above the highest level to which she could climb.

It had become customary, since we had started flying as a

crew on the long transatlantic flights from Bermuda, to let the aircraft go with the wind rather than always to be fiddling with her, altering course.

Sometimes it pays to do this in terms of time for the flight, though the distance flown may be slightly greater. On my first cross-country flights when training in England for the first world war, I was lost if I missed for a moment the sequence of railway lines, forests, villages, and towns by which we picked our way with intense concentration from airfield to airfield. I was unconscious of the air, except that I could see through it to the objects on the ground which I needed to identify to find my way. Wind had little to do with it, and its possible variation in force and direction at different levels nothing at all.

In the course of time, experience, and improvement in the performance of aircraft, I began to think more in terms of the air and less of the ground. In the beginning it was to the earth we went in emergency, striving to cling to its security, and somehow to get down in a field. Now, for a long time, it had been to the air, to put as much of it as possible between the aircraft and the land. Here, over the ocean, the scope was wide, and one had abandoned altogether the original outlook of hopping from twig to twig.

We drove on through the night, with wraiths of cloud smoking over the wing, out to the coal-black holes between these gray shadows that for a few moments enveloped us; and on, riding the bouncing aircraft with the airstream crackling and blasting by the hull, and the roar of the motors hurrying free on the wind.

To keep her down in this broken sea of air became a self-inflicted torture when I knew that by raising the nose and letting her climb a few hundred feet she would cruise serenely over the cloud. But we became resigned to it as a necessity, and I lost most of the physical discomfort in the interest of finding the wind on which the flight had been planned.

As she burned down the fuel in the main tanks, Jock turned on the transfer pump and sent up the supply from the auxiliaries in the hull.

Though the power needed to maintain efficient flight was

slightly higher in the turbulent air, I had been able to reduce it progressively, till at midnight, local time, when we had run 1375 miles from Clipperton Island, she was down to 440 brake-horsepower per engine and burning a total of 63 U.S. gallons per hour, a rate at which the range was creeping up impressively. We still had nearly twelve hundred gallons aboard. At the lower end of the weight scale this would give her a clear twenty-two hours more range, and more if we were really extended. Allowing for another seventeen hundred miles to go, we were in a good position now, and set to reach our destination if we could keep out of head winds.

Soon after midnight Jock reported that the starboard oil pressure was running low when we had not yet run half the distance and were still some eight hundred miles from land of any sort. There was nothing to be done about it, however, so I told him to let me know if any further variation showed on the gauge. I just hoped it would not go on falling.

But it changed the whole aspect of the night, and in my mind it spread below us an ocean, when before there was only the aircraft and a distant island. Before, I had been thinking of fuel consumption and winds, but now the airplane had become a thing to be reckoned with, to be considered in itself instead of being accepted as our means of passage through space. I went back up to my seat in the pilot's cabin, linked up the interphone to keep in touch with Jock; began to consider the possibilities, and be ready for emergency action.

It might be the gauge and not the pressure to the bearings. If so, there would be no rise in oil temperature even if the gauge dropped to zero. But that could be no guarantee that it *was* the gauge.

If the oil temperature rose there would be no doubt. We would have to shut down the engine before it was seriously damaged, feather the propeller, and try to hold on with one engine. But if it didn't rise I would have to gamble on the defect being in the gauge, keep the engine going and hope it didn't burn up and disintegrate.

Or would I? Was it worth the risk, or was it better to shut down the engine now, and use it only as a last resort if she wouldn't make it on one? There might be some indication from the color of the exhaust flame, but that wasn't infallible either. It was a gamble either way. I decided to keep her going if the temperature didn't rise, and run the risk of structural damage for the hoped-for advantages from continuing on both engines. If the pressure held up now and fell later when we were closer in to the Marquesas I would feather the propeller, go in on one, and put her down in the lee of an island and investigate.

In the back of my mind were the parachute flares, if we had to land in the sea before the light came.

I called to Jock, wanting to know what he was seeing on the gauges: "How is she, Jock? Temperature all right?"

"Aye, Skipper. She's holding now, and the temperature's normal."

"Fine, Jock. Let me know if she drops any more."

"I think she'll be all right now."

Jock seemed to know what was happening inside his engines, without the indications of gauges.

Then Hendy's head came in through the bulkhead door. "Not doing any good here, Skipper."

"What's going on, Hendy?"

"We're only making a hundred and five."

"I'll come down and have a look."

He had two star positions on the chart—beautiful three-star fixes, with the lines cutting almost at a point in spite of the bad conditions. The run between showed only 87.5 miles in fifty-two minutes. That was not good enough.

"All right, Hendy, we'll let her down under the cloud and check again with double drifts."

Hendy returned to his chart, meticulously drawing in the vertical section of the weather.

Birks was a dark shadow, silent, up in the starboard pilot's seat.

I called up to him: "Going on down, Norman. See if we can get the easterly again. Just let her down steadily till we're under the cloud."

I went back up, sat there, with the bulkhead door shut to keep out the cabin lights, and turned down the fluorescent instrument lighting for vision in the darkness. At two thousand feet we were still in the cloud layer. At fifteen hundred the ghostly opaque walls were still rushing in and covering us, to slip away and leave us in a black abyss before another appeared ahead.

I shielded my eyes and looked down, searching for the surface of the ocean, but could see nothing. I shone my flashlight on the altimeter to check the barometric pressure setting. We had set it at Clipperton for zero on the lagoon. Here I thought the sea-level pressure would be lower, almost on the equator, but not unusually low because of the normal weather. I guessed at the altimeter setting on this basis, and turned it back, bringing us down about another hundred feet. Soon it was reading a thousand feet as we continued to descend towards the sea.

To fly at this altitude brought the sea very close, and with a doubtful engine, but we had to weigh the demands and supply the strongest, which was the need for avoiding the head wind, even though it was light. We had adequate fuel now for all normal eventualities, but we didn't know that they would be normal right through to Bora Bora. The engine with the doubtful oil pressure had to be accepted without consideration, because the following wind was still more important than height for possible engine failure in the darkness.

At six hundred feet the aircraft was brushing the bottom of the cloud, and a moment later was clear of the base. Birks brought her up to level flight, and I called for Hendy to take the drift.

Now there was a dim suggestion of swiftly passing shadows that told of the surface close below the aircraft, but so vague that no reliable estimate of our height above it could be made. I believed that the altimeter setting could be relied upon to keep us out of the sea down to an indicated height of three hundred feet, but I switched on the landing lights to watch for signs of waves in the beams. They cast a weird diffused light into the air ahead of us, showing nothing. I switched them off and let the darkness close around us again.

Because of the many drift sights which had been taken to check the wind, only four flares remained, but now, at this critical stage, its force and direction at the flight level had to be known, at the expense of later observations before dawn.

On the signal from Hendy we turned her sixty degrees off course, and held her there while he let go a flare and measured the drift.

As he opened the tunnel hatch a blast of smothering sound rushed into the hurrying roar of the airstream and I felt the night suck at the air in the hull. In my mind I could see him back in the tail crouching down behind the opening into the void below, picking up the flare in the drift sight, following it down the bars and swiveling the sight till the light ran true; then, with the aircraft back on course, taking another drift sight and running out the wind on the computor.

Again it was for us; from east, 14 knots. I decided now to keep under the cloud base, whatever the lowness of the altitude, since it was obvious that here the stream of air flowing westward was very shallow—not more than a few hundred feet deep.

Up through the gaps the stars were brilliant, in an active sky suggestive of wind. Up there it would be strong, against us. We would hold on down in the shallow airstream till the first sign of dawn. Then, if the cloud layers were still thin, go up to calm air for good sextant conditions and a fix from which to lay a final course for Fatu Huku.

Through the night Bligh had been in contact with the base aircraft at Clipperton and with Dorval radio direct, and had been able to pick up time signals from Washington and San Francisco to check our chronometers.

I had taken a few sights to obtain the information I needed for checking the fuel consumption in relation to our distance made good, and the groundspeed. Rather than disturb Henderson each time I wanted to know the aircraft's position, I kept a running check on my own chart on the table of the small office on the starboard side. Constantly to interrupt the work of an air navigator in the circumstances in which we were flying is to keep nagging at an artist painting a picture. Some interruption was inevitable, because of the overlapping of our activities in so small a space. But each of us interfered as little as possible with the other.

The chart table had to be used for meals, since there was no other place to lay out the food and the mugs of coffee or fruit juice that we had at odd times. A drawn line separated the end set aside for food from that which was kept for navigation. It was a rigid, unwritten law in the aircraft that nothing would be put on the navigation section of the table, whatever the urge to do so.

We had covered the food end with a sheet of aluminum and kept cloths handy to clean this down, so that it was always kept decent when not being used. At the side was a large cardboard carton in which all refuse was dumped.

These were important details. In *Frigate Bird* we shared an appreciation not only of the need for keeping order and decent conditions in the aircraft, but of those conditions for their own sake. In addition, on any long air voyage respect for individual habits is essential.

Len Bligh had a Hamilton watch which he hung up on a particular piece of cord to a knob on the radio set. To have this watch there was important to him. Jock had special places for some of his engineer's equipment; Birks had a spot over the instrument panel for his pencil and his cigarettes.

There were these small personal niches and angles to life in the aircraft which, though none of us mentioned them or were particularly conscious of them, were subconsciously recognized as part of the person concerned and not to be trespassed upon. They gave to each of us the home which was our castle. A flying team is effective only when each member is an individual, for most are individualists. They work as a team only if they retain that individuality and freedom of action. Each has a different view of his world in the air.

The pilots are out with the night, followed by the dark shadow of the aircraft, stared at by the pale eyes of the luminous instruments which seek confidence from them.

The navigator has light on a chart with a series of marked positions, and these reach out through a process of evolution towards the ultimate attainment of a point, which at the end materializes and becomes a place upon earth. There is nothing between the lighted chart and the stars. The aircraft, if the navigator is not also a pilot, is merely the platform on which he is working.

The radioman lives with sounds, and with fine touch upon his equipment in search of them. They bring to him, as his personal objective, proof of other distant inhabitants in a world that is otherwise empty. He is unaware of the aircraft unless some drastic change breaks the rhythm of its flight.

The engineer, up in the tower, is the central receiver of sound and vibrations, none of which consciously occupy his attention unless they change, when he becomes instantly a sensitive and accurate recording system and a medium for action. He is usually alone with his intimate knowledge of his machine.

Flying low, we held the easterly wind, drumming along in a period of the night when everything seemed to be flowing automatically. In an hour the dawn would begin to dissolve the night behind us, and soon afterwards there would be no stars. The time had come to break the spell, to make us conscious of ourselves, to lift us out from a state in which we were not tuned for the work before the dawn.

I called up Jock and asked him to heat up the stove for coffee. I had now an impression that I was being flown by the airplane; looked across at Birks to see how much he was with us.

We had been flying now for nearly eighteen hours, for most of which Birks had been on watch in the pilot's cabin. Each of us had been alert for this period, though the subconscious dimness of the early hours had lately crept over us.

I saw that Birks was still staring ahead of him, occasionally reaching across to touch the turn control on the auto-pilot to keep her on course. "Like to go below for a while, Norman; have a walk around? There'll be some coffee going in a few minutes."

I was careful not to imply that he might need a rest, because I knew that whatever the need he would not admit it, even to himself. On a flight of this kind we could not work the aircraft on a watch-for-watch basis, because each of us had something to do all the time. I know that Birks had settled himself up in the pilot's cabin, prepared to stay there till we reached Bora Bora, in order to give me freedom of action for other activities. He was reluctant to leave the cabin, regarding it, I think, as a matter of principle that he should stay there. He went below, but I think more as a concession to me than an admission that he needed to.

I needed to fly the aircraft. I slipped out the auto-pilot and took the controls to line myself up with her again. She was cracking along at about six hundred feet on the altimeter, bouncing and swaying enough to shake me back to flying her instead of being flown. In a few minutes I began to feel the movement of the aircraft in the flight instruments, and a sense of rhythm again flowing through me. I could feel in my hands and through my body the sound of the motors flowing on with the aircraft driving freely with me in easy undulations over the waves of air.

Through the perspex of the cabin roof an occasional star looked down, moved on, and was shut out by the cloud. The layer was still thin, with clear night above. It was time to fix our position before dawn faded out the stars. I brought up the power enough to let her eat into some height, and lifted her into the cloud base.

At three thousand feet she was out through the top, flying in a stream of dead calm air.

In a few moments she had settled down to flight so steady that, after the turbulence below, she seemed to have stopped, frozen still in the clear air below heavens so brilliant that it seemed I could have reached out and picked off any star. I trimmed her down on the course and engaged the auto-pilot.

A strong smell of fresh hot coffee came up from the cabin, giving me a snug feeling within the aircraft. Leaving her to the auto-pilot, I took my sextant and drew down a star. I watched it for a while, rising and falling slowly against the dimly lighted bubble as the aircraft swayed slightly on the mechanical control of the auto-pilot. I touched the adjustment till the apparent rise and fall of the star were equal, directly averaging the sights. I took the time on my watch and read off the altitude of the star, Canopus. I took another star to give a wide cut with Canopus, and saluted the heavens for this friendly information.

In a few minutes Birks came back to his seat and I went below to give Henderson the cockpit. He went up for the stars that he wanted. I put away my sextant and put the sight book on my table. The stars could be worked when they were needed.

Bligh was beginning to lose contact now with Dorval and

with the base aircraft. As dawn approached, the signals were becoming weaker, and it was obvious that they would soon fade out. I was glad of this. Now the time was coming when Bligh could rest without loss of face and refresh himself for the later stages when we hoped that some radio contact with central Pacific stations would be possible.

No further change had shown on the engine instruments, so we just kept our fingers crossed and told ourselves the apparent loss of oil pressure must have been in the gauge.

Though there was little action for Jock after the hull-tank fuel had been transferred up to the mains, he had to keep watch on the engine instruments for any variation of temperature or pressure. Also, we had been breaking down the mixture below automatic lean to attain the most economical running. To do this without damaging the engines required care and intelligent manual control.

I sat with him now in the cabin for one of the brief periods when he had descended from his throne in the tower, both of us building in for the next few hours with the coffee he had brewed.

Power was down now to 420 brake-horsepower per engine, and consumption, by my calculations, to fifty-nine U.S. gallons per hour and fifty-eight by the flow meters. I had left her with this power for the last three hours, except for the few minutes of climbing, and if progress proved to be as I expected from the sights, would probably let her go on without further reduction to the end of the flight. We were building up a groundspeed score against falling short of our objective which, if it continued would soon allow us safely to drive her and cut down the duration of the flight.

Hendy came back with his sights while we were having coffee, and started to work them. There was a look of triumph in his eye which made me smile with him as he put down his things on the chart table. He had Fatu Huku in the bag all right—I could see that, and I knew how he felt about it.

From a position in latitude 8° 20′ south, longitude 137° 29′ west, we laid a course for Fatu Huku, 103 miles distant by reckoning, as the dawn of the day following that on which we had last wakened on the lagoon at Clipperton Island was coming over behind us.

That period, and our passage through the night, seemed one which could not be gauged in time. The island was an infinitely distant place, and we were conscious now of approaching another world.

Back again in the pilot's cabin, I saw that the cloud ahead was thinning out to a few scattered wisps; and below, the wind on the surface was still from east. I let her go on down again to a thousand feet, to ride on this wind for the Marquesas. We held her there steady on the course while Henderson checked the drift at the lower level.

Thirty-five minutes after the position had been fixed from the last of the stars, Birks sighted Fatu Huku. We had both been gazing ahead into the slightly hazy air of dawn for the first sight of this island.

Now it was there.

We could see the dim outline of a small, steep island faintly but definitely visible ahead through the starboard panel of the windshield. It appeared as a rounded pinnacle of rock standing distant, ethereal in the soft western light of early morning.

But I saw it as a most material thing. Fatu Huku was a rock in the ocean which clearly stood as a mark symbolizing our victory in the race for range. Whatever the next thousand miles had in store for us, shortage of fuel was no longer a problem.

That was my first reaction to the sighting of this island. It was a practical reaction which, immediately it was recorded in my mind, was followed by a sense of fulfillment more complete than we had experienced from the first sight of Clipperton.

Flight through any night over the ocean is an adventure, and every landfall a discovery. This night flight, through air which had never before heard the sound of engines, gave to this small island a significance greater than it would ever have if we flew to it again. We silently watched Fatu Huku coming in towards us.

Ahead on the port side there soon appeared the outline of another island, dark under a covering of heavy cloud. That would be Hiva Oa. As we approached, flying low at five hundred feet, it was a grim and impressive sight. The ocean was savage and heavy rollers tore at the base of the island, sweeping in to attack the land, bursting on the rocky coast and surging high into dark caverns in the cliffs.

I skated her round the headlands and bays of Hiva Oa, feeling her soar like a seabird. This was movement, after the long, steady progress through the night. We were discovering the new world of the morning.

Ahead, on a bright green grassy slope, something was moving. The land rushed in to us, and we came in over the headland a few feet above the ground. Some wild horses looked up at us, amazed, and swept away for the timber.

I pressed her down for the beach the other side, really felt her flying, laughing on the air as she picked up the beach and ran it smoothly under the wing. Another headland was coming. She left the beach and lay on the air seaward, pressing away the menace of the cliffs.

She floated up in the turn, and I rolled her slowly over to come back at the island. We played with the coastline to the end of Hiva Oa, leaving behind us a sheer mountain which hid its peak in a cloud base of three thousand feet.

We saw no more of the Marquesas Islands, and headed now for Rangiroa, a large atoll on the northwestern end of the Tuamotos, 570 miles from Hiva Oa.

The weather was still holding fine, and the wind west of the Marquesas was dead behind us from northeast.

We had been flying now for twenty hours. With an apparently clear run ahead for Takaroa, the quiet desire for sleep drifted over us as the aircraft settled down again to the smooth progress of its course.

I now felt relaxed and at peace. With the auto-pilot engaged I lay back in the seat and for half an hour let the aircraft carry me on in a detached condition of consciousness, the effect of which is as good as sound sleep. One is never completely unaware of the aircraft, but is certainly so of other external influences. I could have gone below and slept soundly in a bunk; but to do so breaks the thread of flight and erases the picture the mind has painted ahead. By drifting along with the aircraft, up in the pilot's seat, one never loses contact; but at the same time expended resources are restored by the state of semi-oblivion, the dim consciousness of which is itself a sublime peace and therefore an effective energizer.

I was brought back by a plate of bacon and eggs placed in front of me by Jock, who had prepared breakfast when we cleared Hiva Oa. While I ate these I began to think more about the way ahead.

A light pink stain, low in the sky, which had not been visible from the Marquesas now showed signs of weather away in the southwest. That would mean loss of the sun at some stage before Rangiroa, beyond which Bora Bora lay slightly less than three hundred miles. It might mean loss of the sun right through to Bora Bora, and that could be serious.

Leaving Birks with the aircraft, I went below to look into the radio situation with Bligh. If we could use Bora Bora for homing, loss of the sun for navigation would not be serious. He called Bora Bora; but no reply came over the radio.

He kept on calling, and somebody else came in. It was Tutuila, twelve hundred miles farther west, in U.S. Samoa. Tutuila got through to Bora Bora for us, and eventually Bligh managed to contact that station direct, the signals coming in very weakly.

The information we received was not reassuring.

There was no radio range at Bora Bora, no H.F./D.F., and the frequency of the beacon, 1688 kc., was outside that of our loop receiver, the limit of which was 1500. We could not use Bora Bora radio for navigation.

But Bligh was able to get a weather report. It was clear and concise, but it told of conditions which in a short time could have improved to almost perfect weather, or deteriorated to misty rain with very low visibility. All we could do was keep in touch with Bora Bora and go ahead on the basis of using the sun, with very careful checks on drift.

Uncertainty about radio aids at Bora Bora, and the fact that we could not use the beacon, was another legacy from the unstable political atmosphere which preceded the flight. In the circumstances, it had been quite impossible for the signals people to get together and really iron out the question of radio communication and direction. The signals men of both sides were willing and knew what to do, but the constantly changing political front never allowed them to get near each other for long enough to establish a liaison from which some practical plan could be devised and put into effect.

Before we reached Rangiroa the weather was down.

We sighted Manihi and Ahé. They lay flat on the sea a few miles away off the port wing. Soon afterwards the aircraft ran into rain and they were blotted out as we approached the fringe of the immense lagoon of Rangiroa.

We passed on over the Rangiroa lagoon, lying smooth and colorful below us, and identified a point from which to make our departure for Bora Bora. From this we flew on in rain and under low cloud with its base at a thousand feet.

The wind began to swing, changing gradually from northeast, through east, to southeast, constantly varying in force. The changes in force and direction were too frequent to chase them with the compass, so we held on the course from Rangiroa and Henderson kept a running record of how we were fetching up. While he recorded the changing drift, and thus could keep a check on the position, deviation from the course did not matter. That method was more reliable than altering course frequently in an attempt to keep her running dead on the line for Bora Bora.

I had no more inclination to sleep. The flight was narrowing now, closing in so that we had to think about seeing an island two miles away instead of being concerned with a state of progress across an ocean.

I had finished with power and fuel consumption. There was enough fuel in the tanks to reach our objective and to go on flying till the end of the day, beyond which there was little advantage in being able to keep in the air. I had mentally recorded the fact that if the weather should shut down completely ahead and cause us to miss Bora Bora, enough fuel had been conserved to return for the broad front of the Tuamotos and there to reach some lagoon before the night.

Bligh was achieving with the radio all that it could give us, and this was communication with Tutuila and Bora Bora, from which the weather reports began to improve. The cloud was gradually breaking and according to our reckoning, the island was ninety miles off. This started me thinking about the chart for the layout of the lagoon.

Fifty miles before the island was due the weather cleared, and we flew at two thousand feet, cracking along on a good easterly wind with some scattered cloud drifting over the wing.

Twenty-seven hours after we left Clipperton I saw the dim shadow of a high island ahead. I took out the chart again, identified it as Mount Temanu, the great rock that caps the island of Bora Bora. There was no doubt about it. The shape we saw was the solid land of a high island.

Birks and I looked across at each other and I reached for the ignition switch of the starboard engine, turning it onto the port magneto. There was a barely perceptible change in the note of the engines. Then they went on running as they had for three thousand miles, spinning their way in for the island.

I turned round to tell Henderson; found him there, looking ahead through the bulkhead with the navigator's light of victory in his eye. I called Bligh from his radio and we all watched it coming as we had the first sight of Clipperton. I called through the interphone: "Bora Bora, Jock."

"Aye, Skipper."

The "aye" was perceptibly longer than Jock's usual reply. It was packed with meaning.

Five miles off I slipped her off the auto-pilot for the last time, and felt the eagerness of the aircraft. She was sliding down the last slopes of the air. It is always like that coming in at the end.

We saw below us an atoll, with a mountain in the middle of the lagoon. Round the high land of the island, sloping down to palm covered groves at the base of the mountains, were the narrow waters of a lagoon enclosed by the low coral land of the atoll island. I laid her in a steady turn, letting the wingtip sweep slowly round the great rock of Mount Temanu, and looked down for clear water to float her on.

Here was no horror place like the Clipperton lagoon. Deep blue water went down from the central island, shelving up in clear sandy shoals to the ring of small low islands enclosing the lagoon. Some coral heads were scattered in the shallow water, but the deep was clear. The U.S. Army airstrip lay like a long white slab of concrete on the atoll rim, and in a sheltered bay under the mountains we could see the Navy base.

Down in a bay of deep water a launch was circling a buoy, showing us the seaplane anchorage. I lined up a run of deep water close by the bay and brought her round.

She came in low across the outer island, floated quietly over brilliant shallows, and skated onto the blue surface. I heard the pattering of the waves running under the hull, and felt them knocking at the skin—strange contact with the world again, but real when she surged to rest and lay there waiting to be taken away for the mooring.

We picked up the buoy in a calm bay of great beauty and I sat for a moment in the hush when the engines stopped.

The luxury of being in an apparently secure anchorage immediately impressed itself upon me. To be on a mooring with a white buoy was an extraordinary situation.

It was early afternoon when we landed at Bora Bora. We were met at the shore by officers of the United States Navy and Army, who received us warmly and with appreciation for the fact that we had made a long flight and needed rest.

I think we were all feeling well charged with the stimulant of having reached the island, and had no feeling of exhaustion; but it had been a strenuous flight, not so much because of its duration but of the need for concentration on gaining every point, with unknown conditions ahead. Within a few minutes of landing we were driving along the shore on a narrow road through the palms, shaded by their fronds meeting overhead, and cooled by the easterly breeze that had brought us to the island.

Ahead was that delicious period of relaxation which already was beginning to flow through me.

My thoughts were suddenly swept back ten years, to the ride in an enormous automobile, with leis, and motorcycle police shooting ahead with screeching klaxons clearing the traffic into Honolulu. Now I rode in a jeep with the wind in my face, feeling fine. This was the same as arriving at Honolulu on the first flight from Australia to the United States in the Lockheed Altair, when we came in to Wheeler Field in the little single-engined machine at the end of the three thousand mile flight from the Fiji Islands. This time Henderson had been the navigator. But his problems were the same and no easier than mine in the Altair. This was war, that was peace—now there were no trimmings.

Bowling along in the jeep I felt the same welcome from the same kind of people as I had felt that day in Honolulu. Everything we wanted was turned on for us when the jeep swung into the base.

Captain Beattie, the Navy officer commanding the base, had invited me to stay at his quarters, and it was to this haven that I went early in the afternoon to relax and enjoy the full realization that this was Bora Bora. There were the luxuries of a shave, a shower, and a bed, and luscious pineapple beyond which food or drink would have been only irritants to the delicious prospect of rest.

I lay on my bunk and let all the impressions of flight play lightly through my mind, viewing them with a pleasant detachment, till they drifted away and I was aware only of the warm, sweet perfumes of the island as I floated into sleep.

With full tanks again, we flew eastward from Bora Bora to select a refueling base for aircraft which otherwise would have to make the long nonstop flight from Clipperton. We flew back to the Marquesas and made a thorough survey of these islands, which had not been possible on the flight from Clipperton because of the narrow fuel margin. We ranged over the Tuamotos; out to Puka Puka, the farthest eastward of this atoll world, and at night stayed in the sheltered calm of any convenient lagoon.

With the fuel again low we returned to Bora Bora and the next day flew to Tahiti, there to select a site for the future air base. Sailing low on the air round the coast of the island, with the great towering peak of Orophena deep in the trade wind cloud, we saw many places on the reef where a runway could be laid down, but finally decided upon a site at Faa'a near the port of Papeete which was obviously suitable for both flying boats and the construction of a landing strip.

As I write this story there is some satisfaction in knowing that Tahiti is celebrating the opening of the international airport on the site we recommended after that day in 1944. The success of this project, involving a magnificent 10,000-foot strip on the reef at Faa'a, is very largely due to the foresight and persistence of the French pioneer, Colonel Louis Castex.

From Tahiti we flew back to Bora Bora, and the next day left our friends at the U.S. base.

The flight from Bora Bora was mainly a survey of islands at which some air facilities had already existed. We passed westward over Mopelia and the Hervey Islands, and went in to Aitutaki lagoon. It was studded with coral, but shallow water in a small area under Ratuatakura Island lay clear over the familiar greenish yellow of sand. Into this refuge I stall-landed the aircraft, brought her quickly to rest, and anchored immediately. We stayed there overnight, under the lee of the island.

Examination of the lagoon in the morning proved Aitutaki to be suitable for development as the main base west of Bora Bora. I lined up some marks on the land that gave us safe water for the take-off, taxied *Frigate Bird* out of the small cove into which I had dropped her the evening before, and lifted her off the water for Nukualofa, Tonga.

At Nukualofa we knew again the surprising luxury of deep blue water and a mooring, which discreet inquiries convinced me was sound below the newly painted buoy. A day there was enough to look over the landing strips and other facilities which existed.

Flying south for New Zealand, we examined the Kermadecs; took her close in round Raoul Island, where Hendy had once been on a survey job for the New Zealand Government. I let the wing follow round the contour of the island, by the rugged hills and densely wooded valleys over which he had worked his way on the surface years before.

It was a still, clear day, with brilliant colors in the water where the steep sides of the island went down in deep purple to the ocean. Soon Raoul Island had gone, like the rest; and we passed by the other rocks of the Kermadecs.

New Zealand was under cloud. We slid in low under the cloud-base, crept up on Auckland in misty rain, and surprised the slow gray water of the harbor by landing on it.

An enthusiastic welcome by the New Zealand Government

and Air Force warmed us again to the basic purpose of our
venture and to our friends in this country.

We left Auckland for Sydney on a wave.

Ten hours later I saw Australia.

As the dim, blue streak of land crept in over the nose of
Frigate Bird, I saw too the ski shack with red window frames
up in the Laurentian Mountains where my wife and family
were living, and was content with what lay behind us in the
South Pacific: at Clipperton, and through to Bermuda.

We passed in over the coast at Sydney, and I laid her over
in a turn for the airport circuit and called for landing clearance from the control tower.

"JX 275—you are cleared to land."

I sank the aircraft in low over the houses, and slithered her
onto the water of Rose Bay. In a few minutes she was on the
mooring, with the engines stopped. I stood up through the
roof hatch, leaned on the deck there, and felt her relax.

I looked up to the engines, with their ten thousand miles of
ocean flight behind them. They had little to say as they
cooled in the breeze blowing over this secure anchorage. I
thought of their take-offs: of the fifty-two inches they had
given to haul us out of Clipperton, of the clean lift close over
the tall palms of Napuka, of the long night when they had
spun with endless rhythm to Fatu Huku. I heard again their
high-spirited call of wild elation as they blasted their way to
freedom over the Grand Récif, with the hull riding high on
the surface and the savage teeth of coral passing a foot below
the plating. I knew these engines intimately. They were more
real to me than anything around me, except the airplane and
the men with me who had made it possible to bring her
through to Sydney.

At the end of the week in Australia overhaul of *Frigate
Bird* had been completed by Qantas engineers. These men
and women, the government air officials, all at the flying-boat
base, were with us in this enterprise.

The weathered *Frigate Bird* now stood at the head of the
ramp shining silver in the sun. With those who had worked
on her for us I watched her ride down on the beaching chassis and float again on the water. We went aboard and gave

her a shakedown flight. She was singing in tune and ready for
the blue air again.

Early in the morning, we went out on a light westerly for
Nouméa, in New Caledonia. I was impatient to be back and
to learn of the progress at Clipperton Island. We kept her
going, up the route of the Altair, through Suva and Canton
Island, to Pearl Harbor.

At Honolulu we rested. Here was the edge of the new
world again—that swift new world of America. I went into
the sea at Waikiki, swam out beyond the surf, and floated in
the ocean.

The water was cool and refreshing. The sun was low, at
the end of the day, sinking to the horizon out by Barber's
Head. The Pacific was slow, easy, and peaceful; and soft,
changing lights caressed the mountains of Oahu. I moved in
the ocean, easily, just to keep afloat. The gentle touch of the
water took the action from my limbs and the lights on the
mountains smoothed out my mind.

I had been thinking of the objections to the flight, the mo-
tives I believed had inspired them, the possible effect of further trouble, and how it might be overcome.

Here in the ocean I wondered whether it mattered who developed these airways across the South Pacific. It was air
communication that mattered, now for war, and in the future
for the personal contact that could help to bring a better understanding and perhaps a kindlier feeling between the peo-
ples.

I stayed out beyond the surf till it was dark, then took a
wave for the beach. It ran out over deep water that left me in
darkness behind the surge. I came to the surface and swam
slowly in to the shore.

From my bedroom at Halekalani I could see out over the
ocean through the tops of the coconut palms. There was only
the sound of the running surf and the wind in the trees. In
my mind I could hear the music of Hawaii. It was the Pacific. I slept deeply and in peace.

We crossed to the mainland the next night, and went in to
San Diego.

From the Consolidated base we flew over the United
States, landing only on the lake at Fort Worth for fuel and

an evening meal. I felt strangely at home over this continent of North America. There was freedom in the air.

We took *Frigate Bird* out into the black night of the lake and fed her the power for the last take-off. She blasted the darkness with triumphant sound, and left the water below. She swept over the lights of the airport, and I straightened her up on the course for Charleston and the Atlantic.

Beyond the Atlantic coast she bored into an ocean of cloud, and flung ice at us from the propellers. A piece smashed through the skylight, and reminded us that we were still flying. But in the morning she came into Bermuda, and we hauled her back on the ramp at Darrell's Island.

The Dakotas, loaded with equipment, had not yet left for Clipperton. The political ball again was being tossed across the Atlantic. It seemed that the pigs and the gannets and the terns were safe at Clipperton. I could almost feel glad about this, but it was in conflict with the purpose of the flight, and the needs of the R.A.F. in the Pacific. Before any conclusion could be reached in the interminable diplomatic passages which followed, the atom bombs effectively ended the war in the Pacific. But today the pattern of international airlines is spreading to the last empty space, in the Southeast Pacific; and we hear that soon the first big transports will be heading out over the Clipperton track, overflying the island, from Mexico to Tahiti.

Back at Dorval I remembered warrant officer Hicks and recommended him for the George Medal. My recommendation was approved at Transport Command Headquarters; but somewhere up the line beyond that point it must have been turned down, because Hicks did not receive the award. I did not forget this episode and waited for the opportunity to see the C.-in-C. Transport Command about it, but I was posted quite suddenly to fly the R.A.F. transpacific communications service and did not again see Air Marshal Bowhill.

After the war the affair of Hicks and the George Medal stayed in my mind and I decided to do something about it. I remembered that in the British Commonwealth an appeal to the monarch must always receive attention: so I wrote to the

King and appealed to His Majesty for a review of Hicks's case, with the explanation that perhaps my recommendation had not adequately described Hicks's action, that I had now written a book, *Forgotten Island*, about the Clipperton flight, in which the whole incident had been fully recorded; and I sent a copy of *Forgotten Island* with my letter. This, I felt, would enable the subject to be reconsidered without loss of face should the matter come before those who had previously turned down the award.

In reply to my appeal I received a most courteous letter from the Palace, in effect saying that the matter would receive attention, and thanking me for bringing it up.

A few weeks later another letter came, saying that the book had been received and read, and that it was agreed that Hicks's case should and would be reviewed.

Then, after another short period, I had the good news that His Majesty had been graciously pleased to award the George Medal to warrant officer Hicks for his action in saving his aircraft at Clipperton Island.

This was a most heartening end to the affair, not only in knowing that Hicks got what he deserved, but in finding, with a most pleasant personal experience, that the system works.

Chapter 15
The Dollar Bill

In April of 1957 I was traveling by United Airlines from Honolulu for San Francisco when another passenger rose from his seat and came over to speak to me. He opened the conversation by holding out a dollar bill and asking me if I recognized it.

The man was vaguely familiar to me though I could not remember where I had seen him; so I took the bill, examined it, and there across the top was my own signature.

Suddenly it all came back to me. I had signed that bill almost exactly twelve years before, and in the same position in the sky where we now flew in the United D. C. 7: but in very different circumstances.

Early in 1945 I was flying the R.A.F. communications service between San Diego and Sydney and at about midnight one time had turned my aircraft around to return to Honolulu. Soon afterwards, when I came down to the cabin to explain the aircraft's movements to the passengers, one man had asked me to sign his dollar bill, for luck. Here, in almost exactly the same position, five hundred miles out from Honolulu, we had met again; and I was able to tell him the story of that flight.

After I had finished the report on the survey flight with the Catalina *Frigate Bird*, and it had gone on to C.-in-C. Transport Command, I was posted to the communications flight at San Diego for service on the Australia run. This was actually the first British Commonwealth transpacific air service and it had recently been inaugurated with some bright and shining R. Y. 3 aircraft, similar to the Liberator but with a single tail fin, of impressive height, improved performance, and different in the cockpit layout and other details. It was known as the Liberator Express Transport; or, in the U.S. Navy, the Privateer.

I had not flown this airplane, and was destined for a conversion course before taking up the transpacific run.

This was a most sought-after assignment among the communications captains, who were the elite of 45 Group; and there was a good deal of politics going on about it. I was the first Australian to be posted to this service and I did not reach it without encountering some opposition. We had also to qualify for an American airline transport pilot's rating, for IFR clearances under U.S. air traffic control.

The day before I began the special radio range flying course for this rating I was warned by a friend in "Crew Assignments" that the check captain, an American airline pilot, had been approached with a proposal to fail me on the course so that the next in line might get my job. This was a very unpleasant affair and something I had never known in all my experience; but the information came from a reliable source, so I could not ignore it.

When we went out to the aircraft to start the range flying I told the check captain that I knew of this plan to fail me, and that I just wanted him to know I knew about it.

He answered me quite reasonably: "Well, yes. There is a plan to fail you."

The question in my mind was obvious. "What do you propose to do about it?"

"I won't fail you unless you fail yourself; but you're steppin' in fast company, and it's going to be tough."

We reached the aircraft with perfect mutual understanding,

but I wasn't in my best form because I had recently heard from a Montreal doctor that within a few months my wife was to die of cancer, and nothing else seemed to matter very much at that time.

Though I had accepted this diagnosis we did not accept its forecast, and we had eventually found in Dr. Peirce at the Royal Victoria Hospital a kind and wonderful friend and a great radiologist who was able to prolong her life for six years. We had been through all this for a period of over two months and the stark realities of life and death with which we had lived in this time left me quite unimpressed by the petty intrigue which had led to this attempt to fail me on the range course.

My brief reference to this personal disaster is in no proportion to the effect it had upon our lives, and I mention it here only because the need to face up to the enormity of this blow made the attempt to fail me on the range course of little importance. It was, however, rather a hit below the belt at this time, and I firmly decided it was not going to succeed.

As to the "fast company," I wasn't too concerned about that. If "faster company" could ever be found in the flying world than the original team of A.N.A captains, I would be very surprised.

But this American check captain gave me a completely fair deal. He was meticulous and tough, but he was fair, and though the circumstances of our introduction could hardly have been worse, we ended up with a good and, I think, mutually respected relationship.

I finished the U.S. A.T.P. rating with its stress upon the finer details of radio range flying, and was about to enter upon the experience of conversion to another type of aircraft, the R. Y. 3, when I was wakened one night with the news that I was to proceed immediately to take out this aircraft from San Diego for Australia. From the sort of orderly approach to my new job which I liked, it had suddenly flared up into one of extreme urgency, with all the elements for an untidy situation.

I said goodbye to my wife and family, who were to follow me to California, and boarded the Colonial Airlines D. C. 3 for Washington, D.C., where there was an American Airlines connection for San Diego. At Washington airport I ran into Alan Potter from Sydney, about to return for Australia from a war mission to the United States. It transpired that he was traveling by the R.A.F. Transport Command service. When I told him I was taking the run, either from politeness or relief at personally knowing the captain, he seemed quite enthusiastic. I did not disillusion him by telling him I had never flown the type of airplane he was to travel in: nor had Johnny Rayner, also previously destined for a conversion course, whom I had selected as my first officer.

When we arrived at San Diego on the morning of the day the service was due out in the evening for Honolulu I wasn't too concerned about the situation. With a really experienced flight engineer who was familiar with the rather complicated fuel transfer system and other engineering details, I felt that I could handle the airplane after a few hours study of the flight-deck layout, controls, instruments, hydraulics, fuel system, and other basic details.

My first move therefore after reporting at San Diego was to ask for the flight engineer to see me. In due course a smart R.A.F. sergeant engineer appeared and I made what I thought was a purely conventional introduction for the flight before us by saying I supposed he had had a good deal of experience on the R. Y. 3.

The man looked at me with an expression of embarrassed horror on his face and said, "Well, no, sir; I've been on Spitfires all the war and I only reported for duty here this morning."

This really staggered me, because the flight engineer in an aircraft of the R. Y. 3 caliber is an important man. There is not enough fuel in the main tanks for San Diego–Honolulu and one of his duties is to transfer the fuel up from the fuselage tanks, see that he has the distribution correct, and generally to operate a transfer and distribution system which is highly complicated, with many arrows, taps, red lines, and other effects to be properly coordinated. Failure to handle the panel correctly could cut off the fuel supply to the engines, cut out the cross feed, or do a number of things which would be highly embarrassing out over the ocean.

I resisted the natural impulse to register a shock from this latest disclosure of our inexperience and just, I think, managed to put the flight engineer at his ease so that he would not be too worried and could usefully spend the day learning the fuel transfer system, if nothing else.

So, as nonchalantly as possible, I said, "Don't worry. All I want you to know is the fuel transfer system. Be sure you know that, and you can learn the rest as we go."

I will not go into the details of why no other crew was available, but it was just one of those instances which come up in war, when circumstances combine to create an unfavorable situation which cannot be altered. For reasons beyond the control of the C.O. there just wasn't another crew available. I spent most of the day in the airplane, drew up a simple check list, and by evening had acquired some spirit of adventure for this flight.

We had a good meal at the diner, an old railway car restaurant near the airport; and I felt again the familiar detachment from the life around us. Already we were beings apart from the earthy human air of this place with its atmosphere of smoke and food and talk and all the close affairs of life in San Diego. I belonged already with the aircraft awaiting us on the tarmac.

Half an hour before the scheduled time for take-off I received the final weather report and forecast, made out the flight plan and received my clearance from air traffic control. To complete the picture, visibility was virtually zero at Lindbergh Field, with continuous rain from a low overcast. Standing for a moment in front of the lighted terminal, with all the daily routine of life around us, I felt that this was a mad venture.

But aboard the aircraft and on the flight deck it was different. Here was the world to which I really belonged. Around me was something more tangible than all the security of lighted buildings, air-conditioned rooms, restaurants, movie theaters, and even the warmth of homes. While the passengers were coming aboard, I sat in the port pilot's seat and let the full impression of the aircraft work upon me: the sight of the lighted instruments; the glistening engine cowls and propellers outside; the touch of the controls, and the leather of the seat; the smell of oil and metal and hydraulic fluid: and all the significant life system of the great aircraft which was coming alive around me and blending my reactions with hers.

With the loading completed, I started the engines and felt the power within us, charging us with the spirit of the aircraft. I released the brakes and let her move away for the dimly gleaming lights of the runway. I had the feeling that this airplane was on my side; that she wanted to fly, and that provided I made no mistakes in my checks and handling she would go off securely into the night.

We ran up the motors and did the checks in orderly procession, making sure we had missed nothing in the unfamiliar layout. Then I let her roll into the runway and line up for take-off.

Cleared to go, I gradually pressed the throttles forward and the blast of power took charge. The runway lights began to flick by the cockpit window and ahead the line of rain-shrouded flares came in with a rising crescendo of sound and speed. I let her go on, sensing and responding to her needs but letting the airplane do it, as a good one will. A touch back on the tail-trim wheel and the nose wheel came away. She thundered on, lighter now on the earth and nearly ready to go. As the black darkness came in towards us beyond the last of the flares I eased her away and she was airborne in the night.

Onto the gyro horizon, the A.S.I., and the Rate of Climb.

Gear up; and as the speed rose, reduce to full climb power.

As she roared away on the climb, enveloped in the blind night close over the invisible land, I turned her away on the gyro for the sea.

Flaps up. She sagged momentarily. Then moved ahead in freer flight; 2300 and thirty-five inches for the climb.

Rain streamed against the screen with a harsh sound like tiny particles of gravel hosed at the airplane. The life and character of the R. Y. 3 came to me through the seat, the contact of the controls, the flight instruments; and, in the turbulent air, the familiar bouncing, flexible movement of the Davis wing. The diner was gone, in another world. Back in the cabin the passengers were simply going to Australia. Johnny Rayner and I smiled across the cockpit, relaxed now and immersed securely in the night. I flew her by hand as she ate up into the height, seeking the clear night above the cloud.

At eight thousand she broke out through the tops. The last close darkness of the cloud passed in ghostly wraiths and little stabs of swirling air. Climbing still, she rose into the clear, flying now with infinite freedom in a calm and spiritual silence, moving swiftly over the gray sea of cloud, under heavens brilliant with stars and the coal-black depths of space.

As she reached cruising altitude I leveled her off and reduced power. My hand rested on the tail-trim wheel, feeling for perfect balance in horizontal flight. She trimmed well: something slightly more sensitively responsive about her than the Liberator, but alike in the thunder of her power and the strong force of her flight. I looked out to the engines under the cowls, and the wing stealing along behind us in the night.

About six hundred miles out I decided to transfer the fuel. If anything should go wrong with the handling of the system and for some reason we could not get the reserve fuel up to the main tanks for flow to the engines, we would still have range in the mains for return to San Diego. If I let her go on we should not be able to return, or to reach Honolulu. So I passed the word to the engineer to bring up the reserve fuel to the mains.

After he had gone back to the transfer panel I turned my torch on the fuel tank sight gauges on the bulkhead behind my seat and was relieved to see the level rising in the glass tubes. The fuel was going up normally, and I could relax. I switched off the torch, clipped it back on its holder, and settled comfortably in my seat. My eyes were just starting one of those instinctive, peaceful rounds of the gyro, the compass, the temperature and pressure gauges, and other instruments recording the life system of the aircraft when my consciousness was stabbed by a sudden spluttering cough from an engine. My reaction instantaneously flashed to the engine panel and settled on the eccentric flicking of No. 1 engine rev counter. Instinctively I reached to the mixture control and moved it to auto-rich. But at the same time another engine was coughing, and dying spasms of discord were shaking the airplane: in a moment there was complete cessation of life as the whole aircraft seemed to hang suspended without meaning in a night of which it was no longer part. All four engines had stopped, completely.

I cut out the auto-pilot and took her in my hands, letting the nose go down to maintain speed for flight.

"Go down, Johnny, and turn everything off except the four main fuel cocks. Turn them on, and see what happens. Quickly, Johnny. Get cracking."

The last words were unnecessary. Rayner was away.

In the unnatural, suspended silence, I held the airplane as she sank into the night, and into the dark sea of cloud, below which was the ocean. It was an uncanny experience; like living with the death of an airplane. But I really had no thought for anything but to retain control and mentally to press on Johnny's action to get the fuel flowing again to the engines.

In a remarkably short time, though it seemed a timeless age as I sat in the cockpit, life came back to the aircraft. One engine; then another, came in with returning power. I picked them up with the throttles, and soon all four were running again. Johnny came back and we settled her down, tuning in the propellers and lifting her back to the starlit heavens.

In a few minutes we were leveled off again, in clear air, with the aircraft back in her steady orbit. But where did we go from there?

The engineer came back and we looked at the facts of the situation. I did not want to panic into returning to San Diego: but soon a decision would have to be made. Reckoning up the fuel remaining, our distance out, and our estimated groundspeed back, we could safely fly on for another half hour and still return to San Diego.

Tucked into a pocket on the side of the flight deck was an impressive manual on the R. Y. 3, with very good descriptions and illustrations covering all details of the aircraft, including the fuel system. My thoughts turned to this manual because I had studied it myself on the ground at San Diego. So I gave it to Johnny and the engineer to take down to the panel, go into a huddle there, and see if they could make some sense out of all the lines and taps and arrows, and find out how to get the fuel up within half an hour without stopping the engines.

As they left the flight deck, I could not help being amused at the thought of fifteen valuable, high level V.I.P.'s sitting in the main cabin just going to Australia, while the crew were reading an instruction book on how to work the airplane.

It wasn't long before Johnny Rayner returned to the flight deck with his charming and slightly mischievous smile, and reported that the engineer was ready for the fuel transfer. With some concern for the passengers I kept my fingers crossed and passed the word to him to start the vital operation.

All went well. The engineer, in very difficult circumstances, had sorted out his problem and soon had the reserve fuel up for direct supply to the engines. I felt sorry for him because, as I later discovered, other duties had diverted him at San Diego from a full study of the fuel system. But he had adapted himself with first-class initiative to this airplane and did a good job all the way through to Australia.

Though I hadn't much inclination for humorous reflection when the engines stopped, I afterwards recalled to Johnny the story of the well-known B.O.A.C. captain who, in the same circumstances in the night over mid-Atlantic, is reported to have turned to his first officer and remarked, "Dangerously quiet, don't you think?"

So—we ran the lovely, glistening new R. Y. 3 through to Sydney, via Honolulu, Canton Island, Nadi (Fiji Islands), and Auckland, for turn-around two days later on the return service.

But it was not to be. Right on our tail to Sydney was another aircraft from San Diego, flying on some sort of route inspection. It was an ancient C. 87 which was virtually a hack among the new aircraft intended for the regular service. For some obscure reason it was found necessary at Sydney that I should take out the C. 87 for the regular passenger service and the R. Y. 3 would be used for the route survey. Mysterious changes of this kind are always accompanied by tarmac rumors; so, keeping my ear to the ground, I learned that all was not well with this C. 87. Since it was allotted to the service I was to take out, I could not refuse it; but it was my duty to satisfy myself that it was serviceable; so I asked for it to be available for a test flight. This was rather unpopular, but could not be turned down. So Johnny and I climbed up in the rather soiled flight deck and trundled the old C. 87 around on a systematic check flight. We discovered, among more minor defects, a serious leak in the fuel system. I delayed the service twenty-four hours while this was rectified and we made another test flight. The airplane impressed me unfavorably, but I could not legitimately pin down any one point of unserviceability which would justify my rejecting it.

So we lined up for departure on the first stage of the flight, to Auckland.

I observed all the functions of this airplane very closely during the six-hour crossing of the Tasman Sea, and could not locate any one bad thing about it, but I still didn't like it. The performance was sluggish, and at normal cruise power it had a tendency to lose speed, till the nose had to be let down or power increased to restore its equilibrium.

On the run up the Pacific I took very careful checks on fuel consumption, which I began to suspect after the flight to Auckland. These checks showed well above average consumption for the power settings used, and nothing could be done to rectify this along the route. Fuel used on the flight from Canton Island to Honolulu left me with unquestionable confirmation of the consumption figures. The critical flight stage was the last, 2090 nautical miles to San Francisco. The figures of power, speed, and consumption clearly showed that I must have winds for a flight-plan time of 12 hours 35 minutes or better, to leave Honolulu with adequate reserve in the tanks.

By good luck, the forecast winds for the next night gave us exactly this flight time. So we left Honolulu at the scheduled time of 8 P.M.

I instructed the navigator to keep a very careful check from the stars on our distance run, and to keep me informed accordingly. After two hours out it was obvious that we were not doing nearly as well as our flight-plan time. At the end of the third hour winds had slowed us up to a speed which, reckoning ahead, would have put us in the sea about 150 miles short of San Francisco with dry tanks. Winds might have improved to the east and we might have scraped into San Francisco; but I could not rely on this; I turned the airplane

round and returned for Honolulu, where we landed about
two o'clock in the morning.

Soon after heading back for Hickham Field I went below
to explain the situation to the passengers, and just about
where I sat twelve years later in the United D. C. 7 I signed
the dollar bill.

I was very unpopular with our administrative people at
San Diego. But my passengers took a different view and were
kind enough to thank me for not taking the risk of dropping
them in the sea short of the California coast. To be confronted with a new type of airplane is reasonable enough in
war, but there is no flexibility in the rules of fuel reserve on a
transocean passenger service in war or peace.

Soon afterwards, Johnny Rayner was promoted to Captain,
but his luck had run out. He was killed in a Liberator taking
off at night from Sydney airport. The aircraft just didn't seem
to climb away from the airport, but went on level off the
end of the runway and flew through some trees and into a
concrete pipe. Poor Johnny. He was a charming, likeable
chap and a beautiful, sensitive pilot. He had married a lovely
girl from New York, but she couldn't take the life at Dorval,
and had gone back to the bright lights. It had affected him
badly and that was the reason I had asked for him to join me
on the Australian run: new air, new scenes, and new people.
But it just didn't work out as I had hoped.

Chapter 16
Spitfire

THOUGH the habit of seeking beyond the horizon over the ocean had replaced for me the earlier form of flight with light and aerobatic aircraft, there was always the chance that some special incident would momentarily touch the way of the air with a brief but intimate and thrilling experience.

Such an incident occurred soon after the war when I had undertaken to fly the *Southern Cross* for air shots in a film of Kingsford Smith's life. The first take-off in this aircraft, lightly laden, with its high lift, low wing-loading characteristics, was a surprising experience after the thunderous fight for speed down the runway of the overloaded Liberators and the submarine effect of the Catalinas in the first stages of take-off out of Bermuda for Largs with three tons weight above their normal gross. The old *Cross* ran a few yards lightly down the runway and floated into the air, leaving me well behind her in my thoughts and somewhat astonished to find myself airborne.

We had flown to the R.A.A.F. field at Richmond to re-enact the arrival of the Lockheed Altair at Oakland Airport from her transpacific flight in 1934. I was standing with the C.O., admiring a Spitfire resting on the grass by the runway. Of all second world war aircraft, I think the Spitfire is the most inspiring. There is something brilliantly adventurous about her sleek beauty; a deadly menace in her purposeful air of reaching forward to attack, even when resting on the grass.

I was quite preoccupied, absorbed by the beauty of this aircraft, when the C.O. turned to me and asked.

"Would you like to fly it?"

This question quickly alerted me, from the aesthetic enjoyment of the Spitfire's obvious qualities, to a sudden appreciation of my own physical situation. I had not flown a small fighter aircraft since the first world war; and then an entirely different machine, of only 110 horsepower. For some years I had been conditioned to large multi-engined airplanes and to the different psychological approach one develops to them. The small machine is literally flown by the pilot, as a small sailing-boat is quickly and physically controlled by direct contact and reaction. The pilot of a large aircraft exercises a dominant and sympathetic influence over his great machine, usually in an easy, suggestive way, but with sure command behind his suggestion, in much the same way that an elephant man controls his huge but obedient creature whose physical strength is so vastly greater than his own.

Now I was confronted with this light and wonderful aircraft, charged with the terrific power of the Rolls-Royce Merlin engine, not only available to my hand but necessarily to be controlled quite accurately by its movements. I could not, of course, decently escape from the invitation to fly this formidable single-seater fighter. This, and the exciting attraction of being free in the air with all this horsepower in an aircraft which could be so lightly handled, relieved me of any too visible hesitation in accepting the invitation to fly it.

We strolled over to the Spitfire and I climbed into the cockpit, close and intimate with all controls and instruments concentrated in so small a space. Keeping up as easy a manner as I could, I turned and suggestively asked the C.O., standing beside me on the wing,

"Could you show me the knobs?"

"Oh, yes; of course," he replied casually.

Brushing lightly over my ignorance in an attempt to meet his manner without appearing to be too concerned, I nevertheless mentally recorded certain things I positively wanted to know before I unleashed this concentrated source of power and committed myself to control it in the air. There were of course the usual basic functions with any airplane; the things to be checked before take-off, which can kill you suddenly if you neglect them. But raising and lowering the undercarriage and being sure it was locked down before landing: that; and working the sliding hood; and the power combinations of r.p.m. and boost: these also I wanted to know.

With conflicting thoughts urging me excitingly to take off, but warning me calmly to wait and cover every point, I sat in the cockpit checking around this nerve center of the aircraft.

We started the engine, and the Spitfire came to life. My host dropped down from the wing and I was alone with the most terrific impression of latent power I have ever known in an airplane. It was like sitting in the streamlined rear of a powerful engine installation removed from the wing of a large aircraft and placed upon an undercarriage; with small, rigid wings sprouting from the undersides of the cowl. The dominant factor was the Rolls-Royce Merlin with its gigantic propeller. The airplane itself, the Spitfire, with perfect manners and understatement of its own significance, seemed merely the streamlined accompaniment to the engine. It seemed ridiculous that the small throttle lever could control such power.

I waved away the chocks, released the brakes, and gave the airplane sufficient throttle to move away over the grass.

Down at the western end of the airdrome I turned her round, went carefully and deliberately through the run-up and the checks, and lined the aircraft up for take-off. I sat for a quiet moment of mutual anticipation with the Spitfire, then decided to go.

I fed her the power with the throttle and she took me away with violent acceleration, drawing my head back against the rest, and hurrying quickly for flight. I just had time to remember two points I wanted to cover—try to keep her smooth in the very sensitive fore and aft control when the air forces are light on the elevator in the low speed of take-off; and get the gear up quickly immediately she is airborne— when she was away and already heading for her element, the high air.

To leave the gear down too long after take-off is a sure sign, to watchers on the ground, of preoccupation with other too pressing effects: so I remembered to snap it up immediately: but in the unaccustomed movement of doing this, I jerked the sensitive elevator with my other hand on the controls. I mentally blushed at this exhibition of coarse handling, which I knew would be visible, in the aircraft's reaction, from below. But then, with incredible swiftness, the airfield was gone and the earth was sinking away to insignificance in this new expression of flight. I found myself at six thousand feet before I had really tuned in my own reactions at all with this gay, free creature sailing up for the sky with smooth and singing power.

With the hood closed and a familiarizing check around the cockpit disposed of, I began to go with her; and at ten thousand feet I made my first approach to any sort of intimacy. Behind the outward smoothness and good manners was this subtle warning of tremendous power and the violent reaction it would have to any mishandling. This airplane would use up prodigious falls of height if in protest or confusion she headed for the earth.

I laid her into some steady turns and felt her flatten against the air. As I drew her round I felt my own reaction to the turn at high speed in a tightening of my body. I was interested in this personal reaction to a fast fighter aircraft whose endurance of stresses in flight is in some ways beyond that of the pilot. I have never really believed in numerical age as a deterrent to flight of any sort. The more years I live the surer I am that provided the body is given a chance to function normally by supplying it with proper fuel for life, living habits are something approaching those for which the human body was intended, and one spiritually seeks the air and remains lucky in not being stricken by some unexplainable ailment, then there is no numerical age at which the pilot should "give up flying." I have, so far, been lucky in health. I know that I am a better pilot, of any sort of airplane, today than I was ten, twenty, or thirty years ago; and that, unless I lose my health, I shall be better tomorrow than today.

Now, at Five Zero, or fifty, I was intrigued to see what effect this Spitfire would have upon me.

The effect was the most thrilling of any contact I have known with any aircraft. By this I mean my reaction and sense of harmony, inspiration, and satisfaction with this particular airplane in flight; as opposed to the realization of life beyond ours on earth, the divine peace, and the sense of fulfillment which comes to us in long flight over the ocean, particularly at night, in the dawn, and at sunset. These are separate reactions. The first, to the particular airplane as an individual itself; the second, to flight which brings us in contact with and reveals much that, to me, is often hidden under the day-to-day affairs, details, and distractions in the jungle of our civilization.

I found more in the Spitfire than in the small jet fighter I flew ten years later.

After a discreet approach in some long and steady turns I began to feel a response in this airplane; I was no longer just going with her more or less led by the hand, but was coming to level terms with her in the air.

Behind this early confidence was always the warning of power and speed to be observed. This was no leaf on the wind to be lightly flicked in close and playful flight, using little air. Even the earliest impression was of wide and sweeping movement, light and instantly responsive laterally; but now, at more than three hundred miles an hour, soaring in great space-consuming curves of flight with warning resistance to any harsh movement of the elevator. This was an airplane which would, if pressed, respond with strength to the forceful demands of emergency in closer movement, but whose normal desire was for the rhythm of long and easy movement in the sky.

I let her go down with increasing speed in a sense-tightening dive; then began to ease her out, seeking as a matter of interest to reach close to the limits of my own senses. It was soon obvious that this airplane could black me out, but satisfactory to find that I could use, I believe, most of her qualities as a fighter within the limits of my own body. Climbing to the freedom of the high sky I poured her down again with the wild song of speed; drew her up again, laughing, for the sky; and over off the top, to touch the air with momentary surprise in level flight.

This was living.

This was the airplane which inspired John Gillespie Magee, the young Canadian fighter pilot, to write "High Flight," the poem which must forever remain his own inspiring epitaph:

> Oh, I have slipped the surly bonds of Earth,
> And danced the skies on laughter-silvered wings;
> Sunward I've climbed and joined the tumbling mirth
> Of sun-split clouds—and done a hundred things
> You have not dreamed of—wheeled and soared and swung
> High in the sunlit silence. Hov'ring there,
> I've chased the shouting wind along and flung
> My eager craft through footless halls of air.
> Up, up the long delirious, burning blue
> I've topped the wind-swept heights with easy grace,
> Where never lark, or even eagle, flew;
> And while with silent lifting mind I've trod
> The high untrespassed sanctity of Space,
> Put out my hand, and touched the face of God.

And so, no more could be written.

I stayed with the Spitfire for an hour. Then stepped down quietly with her for the Earth.

Down in the circuit, the magic of the high sky stayed only as an impression and a memory. I was setting up another airplane for landing.

But with this free creature I could not think in terms of the familiar transport approach procedure. There could be no "downwind," "base leg," and "final approach" with the Spitfire; but only a natural easy curving descent, finally to line her up for landing.

I watched the smooth green surface of the grass come up to meet her, eased back the stick; power off; she floated a little way, and the Earth received her again.

Back by the hangars, I stepped down from an experience which left me still detached from the surroundings I had left an hour ago, and charged with a sense of fulfillment which remained with me for many days.

Chapter 17

South America, 1951

As LATE as 1951, when regular air services had been established over most of the world's international routes, an important transocean route of the future had never been flown by aircraft. East from Tahiti to the continent of South America the ocean of air over the great empty space of the Southeast Pacific had never known the sound of engines, nor felt the passage of an aircraft. Flight to South America across this ocean had for some time been an enticing challenge to me and to others, but the physical factors of sea and air confronting us had left this flight uncompleted long after the Atlantic, North Pacific, and others had been made successfully, and regular air services had been operating for years.

Though no regular service had yet penetrated even as far east as Tahiti, there were some radio facilities and landing strips (left over from the Pacific war) in the region between the Fiji Islands and French Oceania. It was beyond Tahiti, over the 4500 nautical miles of ocean between the island and the west coast of South America, that no bases or facilities existed. Easter Island, a compulsory refueling stop in this distance, and an important base to be surveyed for the future, presented operational problems. It had neither a prepared landing field nor sheltered water for a flying boat. It was without a harbor and the whole coastline was exposed to the great swells of the southern ocean and to the uncertainties of local weather and sea conditions.

Having thought out many times the possibilities of a flight to South America as a "hands across the sea" gesture from Australia to a people with whom we were some day destined to be connected by air transport, and as a survey of the future route, I finally convinced myself that the factors confronting this flight were acceptable when put in the balance against the effect of the flight if it proved successful. In 1950, knowing that in Sir Thomas White we had a Minister for Air who, as an experienced airman and a man of action, could view clearly the whole picture of such a flight and its purpose, I proposed to the Prime Minister, Mr. Menzies, that we should undertake the flight to South America, using a Catalina flying boat, which at that time was still the most suitable type for such a venture. The Prime Minister gave the project his blessing and passed it to the Minister for Air for his consideration. Since I already knew that Sir Thomas White would support it, my proposal, backed also by Lord Casey and others in the Cabinet, was soon formally accepted and the South American flight project was afloat.

Pressure was immediately brought to bear to kill off the project by disgruntled aviation interests who resented the simplicity of my victory in obtaining government sponsorship for the flight. It was freely said, and circulated even as far out as Tahiti, that the flight could not succeed because of the operational hazards beyond Tahiti, and that if the flight were possible it would already have been done. I was not concerned about all this because I knew the government's decision was firm; but it did introduce a note of further obligation upon me not to fail in this venture.

I selected a surplus but really good Catalina from the Royal Australian Air Force, and, as a further gesture which I very greatly appreciated, the Prime Minister presented this aircraft to me. This completely sealed off any possible consideration of failure. We had to succeed.

I named the aircraft *Frigate Bird II*, the next in her line from the Mexico-Australian exploratory flight in 1944. She was to be our aircraft for the most critical and most demanding flight I had ever undertaken.

With me in *Frigate Bird II* were Captain G. H. Purvis, A.F.C., as first officer; Mr. L. L'Huillier as engineer; Mr. A. Allison as radio officer; and Mr. Jack Percival as executive officer and official correspondent. We had all been associated before and Harry Purvis, Blue L'Huillier, Angus Allison and Jack Percival had special qualifications, the freak qualifications actually which were necessary for the success of such a flight.

I had not seen much of Jack Percival since the Indian Ocean flight in 1939 because our ways had parted then on the natural courses of our personal activities. He had become a war correspondent for the *Sydney Morning Herald* and the *London News Chronicle* and after the spread of war to the Pacific he was attached to General MacArthur's Headquarters. Tenacious as usual to run down his story till he had it firmly in the bag, he stayed on in the Philippines after the last opportunity to fly out of this trap, and so was taken prisoner on Corregidor. He was locked up there in a prison camp for more than three years and was appointed leader in the camp by his fellow prisoners.

Now, for the South American flight, I located him in Korea where he was again serving as a war correspondent. The inducement I offered him of a flight into new air, for which certain failure had been predicted, and my personal invitation again to join me in an uncertain venture, was enough to touch off the spark which brought him out of Korea and down again to the South Pacific.

After the most careful preparation and organization of all we could humanly consider, one factor remained which was a serious hazard: the open ocean operation at Easter Island, and how to get a shockingly overloaded Catalina out of the big swells and into the air for Valparaiso. The nearest answer we could get was the discovery of some JATO rockets in Australia. With the aid of Sir Thomas White we secured these, attached them to the sides of the hull, and had some really effective extra thrust for take-off. We did a practice rocket take-off before departure, my first experience in handling an aircraft in these circumstances. The result was impressive, as the Cat was projected into the air like a fighter.

We set out from Australia, leaving behind us all the confusion of human actions and reactions which pass so completely and are replaced by the tranquillity and vivid life of a Catalina in flight across an ocean. *Frigate Bird II* sailed steadily on by New Caledonia, the Fiji Islands, Samoa, Aitutaki, to Tahiti.

From Tahiti, in spite of predictions to the contrary, we flew on southeast to the legendary island of Mangareva, and, after refueling, into the night for Easter Island. In the early morning we were, according to my astronavigation through the night, approaching that lonely island.

As the first sign of daylight came into the east over the cloud top I drew down Vega to the sextant for a final check on course and it showed *Frigate Bird* now to be six miles north of the track. A sextant check on distance run put her 310 miles from Easter Island, in latitude 26° 05′ South, longitude 115° 32′ West. I gave Harry the course of 084 degrees Compass and tidied up the chart table for the next phase, that of our approach to the island.

Flying above cloud, with the stars faded out and the sun not yet above the horizon, I let her go on into the dawn. There was a mental silence now in the aircraft; the low at the end of the night. Everything about us—the aircraft, the engines, the fittings and equipment in the hull, the master compass on the chart table, ourselves and all the familiar things around us—joined in a rhythm of hypnotic sound that seemed forever to have been our world. I went up to the cockpit where Harry was forcing himself to keep awake after a night behind luminous instruments and stars above the bow.

I called across the cockpit, "Like to have a walk around, Harry? I'll fly her for a while."

"I'm all right, Skipper; but I'll go down for a few minutes."

It jerked us out of the stream. I reached up and disengaged the auto-pilot; felt movement and reaction again through the control column and the rudder bar. I reached forward and twisted the head of the ventilator control. A new sound came in, with fresh air from the dawn. Very soon I was revived and alert again, going with the aircraft, skating close above the white layer of cloud that covered the ocean.

Jack's head came through the doorway in the bulkhead.

"Like some coffee, and bacon and eggs?" he said. "I'll have some going in a minute."

"Thanks, Jack. That would really set me up. Best Harry have his below. Then I'll come down. Ask Angus to let me have the folder with the radio details, would you? I want to check on Easter Island."

Jack's head ducked back and in a moment he passed me up the folder. I trimmed the aircraft to level flight, lined up the indices on the auto-pilot, and engaged the lever in the cockpit roof.

There was a full page of information on Easter Island. Under "Radio Facilities" I read:

Communications: Easter Island provides communication.
        Call sign: CCY
        Transmits/Receives 4335 kcs (night)
        "   "   11400 " (day)

Note: this station works with the mainland at 0130 G.M.T. daily for weather and routines on 11400 kcs.

I ran through that. Angus had the island, on communication. It was the beacon information below that interested me now.

Radio Aids: M/F Homing Beacon—500 kcs.
        Identification: CCY (Manual)
        Type of Transmission: A.1.
        (Range in worst conditions—300 miles)
        No D/F Facilities.

Three hundred miles. We were inside that range now. But bad conditions around dawn. Might be worth trying the radio compass. See what happens.

I reached up to the panel and switched on the compass and inverter; turned for maximum, and watched for a reaction on the needle.

Nothing happened.

Too far out, anyhow, for dawn conditions. I switched off the compass. Harry had finished his breakfast and was un-clipping the night screens that shut the glare from the spot-on the navigation table. He stowed the screens and came back to take over the watch in the cockpit.

Jack had produced a really professional breakfast, which we ate together on the end of the table reserved for this pur-pose.

We both were cynical about the radio beacon. "Do you think it'll work, Jack?"

"No," he said, with a chuckle of conviction.

We didn't say anything to Angus because it was his depart-ment. He always pulled something out of the air, but whether the beacon worked or not could not be influenced by his radio wizardry. It was a purely mechanical contrivance, sub-ject nevertheless, in my experience, to unpredictable tempera-mental behavior.

With the last star position tucked securely away, and a sky ahead that would obviously give us the sun for our distance out, I wasn't much concerned about the beacon. The only thing was this overcast below us. If that held far into the east we should soon have to descend below it for drift sights; and that would cut out the sun. To use the sun we should have to climb up through the cloud again; and that was untidy and meant beating the engines around. I decided to stay on top where she was swinging along freely with the power well down, and work the sun even if it proved necessary to turn northeast, then run down a line to the island. Weather at the island, Angus reported from the radio, was broken cloud, with "average" visibility. The unknown was, how far did the overcast extend before the broken cloud?

The answer came soon, from Harry. I saw his hand come round from the cockpit to attract attention. "Looks like some breaks ahead," he called.

I stood up to look forward through the pilot's screen. Some shadowed patches ahead broke the even white of the cloud top, and far in the distance scattered cumulus towers rose against the morning sky.

I could see it was going to work out. Soon we'd have sight of the ocean below, and the sun high enough up to clear the unknown light refraction. I went back again to sort out approach to the island. There was a slip on the table from Blue, giving me the latest fuel situation:

Fuel used	780 gallons
Fuel remaining	720 gallons
Present consumption by flowmeters	59 galls/hour*

I checked these against my calculated powers and consumptions and found they tallied closely with Blue's observed figures. She was using now 510 horsepower per engine and still had enough fuel aboard to fly to the end of the day. The plan for the flight was working into a good situation in which, if I hadn't lived with aircraft for thirty-five years, I would have admitted that we had Easter Island in the bag.

The chart of the island showed in stark reality the fact that there was no possibility of flat water for landing and no sheltered anchorage for the aircraft. It was a straight-out open ocean situation, for which we had made every possible provision in good ground tackle to hold the aircraft, and in rockets to project her out of the seas to the air when we had refueled. I had stressed the need for the quickest fueling arrangements so that the aircraft would be exposed to ocean conditions and weather changes for as short a time as possible. I reckoned to be on the water only for about two hours. Then to fly out of the precarious position there and let the aircraft loaf along through the rest of the day, and the night, for South America. It was all organized, in my mind. But as we flew towards Easter Island that morning, we flew for the place where failure and disaster had been freely predicted for us; for an exposed island to which there was no alternative even within the long range of *Frigate Bird*. We were finally committed to this island, whatever the conditions happened to be when we found it in this loneliest region of the South Pacific. It tightened me up a bit, but there was something exhilarating about it; an expectation that was fresh with the surprise of impending discovery.

An early surprise came to our sight as the aircraft passed over the first gap in the cloud. There was a complete wind

---

* Imperial gallons.

change on the surface. Instead of the easterlies that had prevailed for two thousand miles from Tahiti, the wind was
coming in from west. Whitecaps on the water, regular wind
streaks, and the sunlit surface alight with sparkling movement
indicated a wind of about twenty knots from 280 degrees. I
suspected that the change had come when the cloud had shut
in below us. That meant that we should be nearer the island
then I had reckoned with an estimated head wind from east.

The sun was showing some altitude above the horizon now.
It was still low, but I needed information to fix the distance
we had run on this westerly wind. An error in refraction of
two or three miles was not important in the big margin that
might now exist between our estimated and true distance run.

The sextant drew down the sun to the bubble horizon to
record an altitude of 12° 50'. When I worked the sight it
showed us to be fifty-eight miles ahead of the estimated position; and now, at seven-fifteen in the morning, only 117 miles
from Easter Island.

I thought also about the drift I had been allowing for
the easterly since the last drift sight, two degrees to starboard.
That would have to be applied back now to the westerly, to
see where we fetched up. Laid down on the chart, it took us
to a point south of the track that needed three degrees off the
compass course to put us in to the island. I made this correction to the true course; then went aft for a drift sight on the
westerly through the tunnel hatch.

The drift was zero. The wind was dead behind us, lifting
her along at an extra twenty knots for the invisible island. I
gave Harry the compass course, and took another sun sight
to prove the first. It tallied all right. There was nothing else
to be done. Thirteen hours of work by us all added up to this
moment; with the aircraft cracking along in smooth air, now
at nine thousand feet to clear the scattered cumulus tops; a
faint blue haze over the ocean, and ahead in our minds the
island of mystery which had been my critical objective for
more than a year. I packed up the navigation equipment, secured the sextant and chronometer for the possibility of a
rough landing, and passed the E.T.A. to Angus as 1533z.*

---

* Greenwich Mean time.

Then I went forward to the cockpit with the folded chart of
the island.

I called up Blue on the intercom. The time was 1458
G.M.T.; 7:38 A.M. at the aircraft. I did a bit of rough estimation and gave him something to look out for.

"Look out for Easter Island in about twenty minutes, Blue.
You should see it ahead under the cloud."

Jack came up and watched through the open bulkhead
while I again switched on the radio and tuned the compass
receiver on 500 kcs. We all concentrated on the needle in the
dial on the instrument panel, and waited. To my surprise, it
started to move, to grope its way round the dial. There was
life in it. Something was happening. According to my reckoning the aircraft was headed dead on the center of the island.
The needle was swaying across zero. It narrowed for a moment, sweeping through about twenty degrees. Then it suddenly
started on a persistent journey round the dial as though it had
decided on some other objective. It never settled on anything
definite again and gave no indication whatsoever of direction
to the Easter Island station. I switched it off, to avoid confusion, and waited for the island.

I knew this was a critical moment for Harry. He had, I
think, acquired some confidence in my navigation; but his
whole background was in the expert use of radio ranges and
beacons that worked, and I know that he wanted that Easter
Island beacon to come in with a good snap on the radio compass needle, to bring it quivering to rest, dead steady on the
bearing. I looked sideways, not too obviously, to get his reaction. He was sitting poker-faced, expressionless, gazing into
the distance. Not bad, I thought, because I knew he wasn't
feeling that way. Jack went back, and I saw him round the
corner, intent, typing a message for the news.

Angus tried the radio compass, but it refused to do anything positive, even for him. So we finally switched it off for
good.

With the island coming in, and now only fifty miles ahead,
I took over from Harry to convert my reactions from navigator to pilot.

It was time to start the descent, for vision ahead below the cloud; so I eased down the power and started her down at four hundred feet a minute. Cloud went smoking over the wing and stabbed her with turbulence in the swirling air of the cumulus. As the minutes passed I disliked even these brief periods without vision ahead. It seemed that we might pass the island without knowing it, and fly on into blank air over the Pacific. I knew from the last sun line that it couldn't be there yet, but imagination kept bringing it closer as we flew through these clouds.

I looked at my watch. It was sixteen minutes past the hour now. The island must be coming in. We broke through a large cumulus to air that was clear in the near distance, with lower, scattered cloud. Behind it were more castles of cumulus. Down now at four thousand feet, and still descending, there was good vision ahead; but no island. I started to search, wide across the blue-misted horizon below the cloud.

There was a faintly darker patch ahead.

Then I saw it; suddenly, and very clearly, outlined against the sky and the distant cloud. It looked like the south end of the island, ending in a clear horizon on the sea. Off the port bow a light shower of rain hung a gray curtain to the sea.

Harry saw it; and Jack, looking over our shoulders from behind the bulkhead. Angus came up, too, and I called to Blue, whose vision directly ahead was obscured by the center section of the hull.

"The island's there, Blue. Looks like about twenty-five miles off yet. I'll turn her so you can see it."

The Cat leaned into a slow turn and Blue saw it too. Then I brought her back on the course.

The blue-hazed outline slowly sharpened as it came in towards us. It brought up an island which was destination for the aircraft, but a place which, now that it was visible there above the ocean ahead of us, seemed more remote from the world than we had ever before imagined. I looked down at the westerly, now blowing over the surface at about thirty knots, and began to speculate about the landing, which I had hoped would be possible on the west side off the village of Hanga Roa.

We came in at two thousand feet, and soon the whole outline of Easter Island was before us, with the rain curtain passing to the north. Then it was possible to identify the individual hills. Rana Kao volcano off the starboard bow, Mount Tuutapu ahead, and to port Mount Terevaka sloping down to North Cape . . .

But the sea was rough for landing off Hanga Roa. Already I could see the surf of the ocean swell breaking there on the rocks, and the surface was broken, with short seas that would batter the hull on landing. We would have to look at the east side under the present lee of the island. A mooring had been put down off Hanga Roa; but mooring was the next thing, after a safe landing.

Thirteen hours and sixteen minutes after we had left the anchorage at Mangareva, *Frigate Bird* swept in over the grassy hills of Easter Island. We looked down to this strange place and found something sinister about its rock-bound coast of cliffs with the sea breaking at their base, and the gaping craters of extinct volcanoes; but between these memorials to a past age of upheaval and smoking lava were peaceful hills where sheep grazed, and we saw several people on ponies. Some fields were cultivated in rich soil, and houses nestled among the few trees that we could see on the island.

We circled the village of Hanga Roa. Angus tried to make radiophone contact with the island. I wanted to know about the fueling, which was my single objective after landing and anchoring in some suitable situation. He called repeatedly, but there was no answer, so I had to decide on an alighting area and go in.

We could see the mooring buoy in the active sea off Hanga Roa, and I imagined the fuel would be somewhere there where a slightly sheltered small-boat basin was tucked in behind rocks and the brown stains of surging kelp. Unless somebody had already taken the fuel to the lee side of the island, we could make the quickest getaway by landing off this west side.

I slid her down from five hundred feet, close over the surface, and made a run upwind at about twenty feet above the sea. The surface, which had appeared doubtful from above, was now obviously disastrous for a landing. Broken water, excited by the wind, roughened the swells with waves that would almost certainly have damaged the aircraft, even from a completely stalled landing. It was a sea in which we might possibly have got away with a landing in desperate emergency, but not a place to attempt a voluntary landing with *Frigate Bird*.

The old rule came into my mind. "Do the thing that is immediately before you." It was the landing that mattered now. Then mooring. Then refueling. I picked her up with the power and soared her up to a thousand feet, turning on the climb to come back over the island.

The cone of Rana Kao slid by the starboard wing. *Frigate Bird* flew over the cliffs of the southeast coast, and the sea was below again.

This was the place for landing. I could see that immediately. A big swell swept round the southern corner of the island and rolled on in endless, lazy movement up the southeast coast. Out under the port wing it burst against the rocks at a point which, with the sweep of the coast southward, formed a half-moon bight in the island that was sheltered and relatively calm. The swell was breaking at the base of the cliffs, but the surface as we saw it from a thousand feet appeared to be smooth, with the westerly, intercepted by the island, released again and hurrying over the surface in furtive squalls for escape to sea. The wind, generally, was coming out from the cliffs and running along the swells. I could see one narrow stretch of sea where the bursts of wind were fairly regular, coming from the lowest region of the cliffs. It was all fairly clear now. If the surface looked possible from a low inspection I must bring her in upwind, along the swell, and touch down as far in as possible, to stop before the cliffs.

I felt free now, and part of the aircraft, as she lay over in a descending turn and swept along below the cliffs. She needed holding, where the westerly hit her in shock waves from off the land; but it was exciting flying, with the sweeping rhythm of a light and powerful aircraft. The Cat had taken off her seven-league boots and was really dancing for Easter Island.

I ran her down the long swells, out from the land; drew her up in a climbing turn and let her fall away again to come in sweeping low over the sea. The surface wasn't as good from twenty feet as it seemed from a thousand, and the long swells were big, and warned of the relentless might of the ocean. I smelt the tang of the kelp and the air of another element; of rocks and cliffs and sea that seemed to forbid the aircraft to enter a region where she didn't belong. Then the cliffs were coming in quickly towards her. I pressed the throttles forward and drew her up, and away. This was it. It wasn't ideal; but I believed she could take this surface, and the swells were long enough for a "floats down" landing. With Blue on the floats control I would have landed with the floats up on a shorter swell, dropping them at the very last moment as she lost air control; but here I judged that the seas were long enough not to catch a float on the landing. I turned across the cockpit.

"Going in now, Harry. Looks good here under the cliffs."

I called up Blue. "I'm going to land now, Blue. Better fasten your seat belt."

"O.K., Skipper. All set for landing."

I touched up the signals for "floats down" and "auto-rich," and took her away downwind from the island.

"Better have twenty-three hundred on the approach, Harry; just in case we need to pull out and go round again."

Harry's reliable hand went up to the propeller controls and he pressed them forward. The engines lifted their voices to a high-pitched urgent note and sang in harmony as Harry tuned them in and clamped the tension on the lock.

I turned her in well away from the land, eased down the power and let her settle to a long, straight approach. I wanted to line up the movement of the swells, try to touch her down as one came under her, and run it down as she slowed on the water. The wind was coming well off the land, fairly steadily in this one lane of approach towards some rocks I had marked in my mind from above. She came in steadily up the lane of wind. The big swells rolled slowly under her; rising, sinking away, and reaching up again as she passed in for the land. I leveled her off a few feet over the tops and pressed on a little power to hold her there, and waited. The cliffs were coming in. It had to be soon. The slope of a big, long swell rose into the corner of my vision. I drew off the power and held her. Then she took it near the crest. The ocean gently touched her keel, sizzling against the metal skin of her hull.

She held it, sinking lightly down the slope of the passing swell. In the trough she had finished with the air. The ocean had her, receiving her gladly now to the blue, transparent water of the sunlit island. The motors panted quietly above us as I let her go on, to feel her way in for water where Angus could reach the bottom with our anchor line.

It had all been easy; a normal landing on the water. But it had left me with an impression of impending insecurity for the aircraft. There was something deceptive about this calm; about the ease with which she slid quietly over the heaving surface of the ocean where shafts of sunlight plunged into its mysterious depths. She didn't belong here, where cliffs hung above her and surf broke in foam and spray on a rocky coast with two thousand miles of open ocean behind the tail. Her safety was absolutely dependent upon the mood of the ocean. I didn't like it. I wanted fuel in the tanks immediately and to get out of this place.

I slid back the roof hatch, stood up on the seat, and, reaching down to the throttles, kept her moving as we watched ahead for signs of the bottom.

Two hundred yards out from the rocks, patterns of the bottom began to appear in the deep blue water. As we moved on in I could see huge boulders and deep holes, kelp and hard bottom. I wanted sand if we could find it—to get our anchor in; and out.

Off the port bow was a lighter patch ahead, but near it the swell was lifting up and tending to break on the sinister brown of rocks close below the surface; but it looked as though there would be swinging room there over what I could see now was a small patch of sandy bottom. This seemed like the best anchorage; only about 150 yards out from the base of the cliffs; but it was clear of the bombora where the swell was lifting, though close to it, and it was the only patch of sand within sight. There appeared to be about thirty feet of water over the sand.

Angus was ready in the bow, with the anchor on the heavy line. I indicated the spot where I wanted to bite the anchor in, and let the aircraft idle for it. A few yards short of the patch I signaled Blue, "Stop engines." The propellers fell over the last few compressions with the final click of the valves, and *Frigate Bird* floated quietly to rest over the light-green patch.

"All right, Angus. Let go the anchor."

He passed it over the side. The shining flukes of stainless steel went glinting for the bottom; down, flashing in the rays of sunlight. It was deeper than I expected; clearer even than the lagoons of the atolls, here in the crystal ocean far from the brown rivers and earth of any continent.

We could see the anchor, still on the bottom. The aircraft drifted away on the wind, slid quietly back till the slack in the line tightened. Angus had a turn round the bollard. He eased it as she brought up; then tightened the new manila down, and *Frigate Bird* was secure.

It was silent on the water, but behind this silence was the menacing echo of surf from the rocky coast. Two riders on ponies were silhouetted against the sky on top of the cliffs, and we could see brown men on the rocks, one of whom had already dived into the sea and was swimming out for the aircraft. He swam easily through the clear water, carrying some object in one hand. We all stood up and watched. More people appeared on top of the cliffs, and some were now picking their way down a track to the rocks below, where most of the men who had first stood there were diving in and making their way out of the broken water, to follow the lone man who was now approaching the aircraft.

He reached the bow and threw back his head with a flashing smile, casting the water from his eyes. He carried a broad sword carved from some timber of the island, which he waved in a gesture of greeting. Angus reached down to help him up and in a moment he stood on the sponson: a glistening, amphibious being shining in the morning sun.

He passed me the sword as a gift and I raised my hand in greeting to him, since it was obvious that we had no common language. He stood on the sponson smiling while I examined his gift. It was beautifully carved, in the shape of a broad scimitar, from some hard, brown timber. The top of the hand grip was the head of a seabird with a hooked beak, and it met the blade in the head of some savage man with a hooked nose and a malevolent grin exposing tightly clenched teeth.

His eye was black, of some material like glass, and it stared threateningly from a circle of brass. Long ornaments were shaped to hang from the ears, and the chin protruded with the form of a tufted beard which completed a face of incredibly sinister appearance.

On the blade a creature was carved, with the body of a shark, wings, and the same head of a hunting bird that surmounted the handle. There was obviously some significance in this head. Here was a point of contact with the man of the island. I held up the sword for him and pointed inquiringly to the seabird's head. In a deep, musical voice he answered, "Manutara."

"Manutara?"

He reached up and moved his hand across the sky, stopped, and pointed across the hills. Perhaps it meant a high-flying bird of the island.

Other men had reached the aircraft now, and were calling to us from the water. Soon about ten of them stood on the bow of the aircraft, sleek and dripping water from the ocean; excited and making to us a flow of unintelligible comments. Some had brought other gifts of carved wood. Most of these were human figures, nearly all with the same type of head, which appeared to me to be similar in features to some of the recently discovered people in the interior of New Guinea. They also resembled the usual representation of ancient Egyptians; and, from the other direction, people of the distant past from Central and South America. This appeared to me to add fuel to the fires of controversy concerning the allegedly single direction from which the people of Easter Island, and of Polynesia, had come in the original migration. I wondered, as I always have, whether they could have come from both east and west.

For more than half an hour after the island men had reached the aircraft there was no sign of a boat with our fuel; nor was there any evidence that boats existed at all on this side of the island. From sign language I had, however, gained the impression that boats would come round from Hanga Piko, the tiny haven we had seen from the air. So we just had to wait till the fuel arrived. I felt frustrated and unhappy about the delay, but so long as the westerly held, giving us shelter under the island, the aircraft was secure.

We amused ourselves, and the islanders, by attempts at conversation, and sought some idea of what we might give them in return for the gifts they had brought out from the shore. Cigarettes and clothes were the answer. I had brought a plentiful supply of cigarettes for the crew and for gifts to people at the islands. So we opened the cartons and distributed them around. For the leader who had come with the sword I found a shirt and a pair of shorts. There was no patronizing charity about this. Easter Island is completely isolated. Very occasionally a ship visits it from Chile. Clothes wear out, and they are a practical necessity, within the simple needs of the place. The gifts we were able to give were very popular, and we were fascinated by the carved figures given to us. They all had the stamp of authenticity which genuinely identified them with this wild and lonely island of the Passover. Here we found it known as Isla de Pascua. To Polynesia it is Rapa Nui. To us it had been known in the past as Easter Island. I think of it now as Isla de Pascua, for that was the name it somehow earned when we were there and when we spoke of it afterwards with our friends in Chile.

About an hour after the aircraft had anchored in this bend in the coastline, shown on the chart as Ovahe Cove, we saw a small boat round the south end of the island.

As the boat approached I could see that it did not have our fuel. This was beginning to be serious. Away in the southeast, low on the horizon, a line of cloud had appeared. It looked harmless enough, but it had crept up over the horizon from that direction, and only an airflow from southeast could account for that. At our anchorage, however, convincing puffs from the westerly still blew onto the water and hurried away to sea. Except for this suggestion of low cloud on the horizon, the sky was clear now, and the island sleepy in the sun and the surge of the surf. I noticed that the swell had a more definite southerly component. There was no sense of security in this peace.

To my relief, it was obvious that the boat was well handled. She came up to the aircraft and I signaled them to let down alongside our bow, which was well in the water. Till they realized how easily the aircraft could be damaged it was better to keep the boat away from the blister, where the nose might go under the tail and punch a hole in the hull on a rising sea.

Leaving Angus and Blue aboard, Harry, Jack and I jumped into the boat immediately it was alongside and I signaled the helmsman to clear away from the aircraft. There wasn't much ceremony about it, but the safety of *Frigate Bird* was absolute priority over all other considerations, and a motorboat alongside a flying boat in the open sea is always a danger to the thin skin of the hull.

We exchanged greetings with the crew of the motorboat, but still had no mutual language for conversation. The helmsman pointed to the shore where a number of people had gathered below the cliffs, and headed the boat in that direction. As we turned in by some high rocks I was surprised to see a sheltered corner, big enough for a small boat to come alongside a perfect natural wall of rock, where the swell quietly rose and fell without harm to the boat. The helmsman placed his craft gently alongside and we climbed out and up to a flat rock where several men stood to greet us.

The sudden change was extraordinary. Out on the aircraft we saw menace in the cliffs of the island; *Frigate Bird* was our home; security, and the island a passing thing.

Now we stood on a warm rock, with people in sunny clothes around us, and contact with earth at our feet. I was brought to a halt by this sudden transition. For a moment there, everything stood still. Then I found myself taking the hand of a man who welcomed us to the island in English. He was the Administrator of the Isla de Pascua. I introduced Harry and Jack. Others came forward; and in a moment we were gathered in by the new and warm human contact of our welcome to Easter Island. The oil company's agent was there. Our fuel was in drums at Hanga Piko on the other side of the island. It still had to be loaded aboard boats and brought round, about six miles, to Ovahe Cove. I arranged with the agent to drive me over in a jeep to Hanga Piko and start something moving about the fuel. Already the time was ten o'clock. At least another two or three hours till it could be loaded and brought round the island. Another hour or more to send it up to the tanks with our electric transfer pump. It would be well on in the afternoon before we could possibly be ready for take-off. Arrangements had been made for our reception. A luncheon. Drive over the island. Everything was set for a leisurely day at this place where the aircraft hung precariously on her anchor. I could see her swinging lightly to the wind, a delicate thing entirely dominated by the cliffs and the ocean; and though the day was sunny and perfect I knew already from that distant, crouching cloud on the horizon that we couldn't rely on the westerly for more than a few hours at most.

We picked our way up the track from the rocks, among incredulous, smiling people, till we came to the top where warm grasslands swept away over the island and people on ponies waited to greet us.

It was all so deceptive; so peaceful and safe; so difficult, I am sure, for them to understand why I couldn't relax and accept the island's hospitality.

We drove in the jeep over grassy paddocks, and came to a long strip of level land where the Chilean Catalina amphibian had landed at the end of its flight from Quintero. It was there under the trees where we had briefly sighted it as we passed over the island. The wing was damaged, like a bird that had been shot. On the nose of the aircraft in bold and flowing lettering was the name, *Manutara*. She was wounded, but proud and defiant.

Manutara. The bird on the sword? I asked the men in the jeep.

They told me *Manutara* was a seabird; high-flying, swift, with sharp, swept-back wings. There was a legend at the Isla de Pascua. The bird flew to the island from some unknown base, laid one egg, and died. *Manutara* never left the island alive. So, you see—*manutara*. And they pointed to the aircraft with the broken wing.

We drove on, down a track to Hanga Piko. The fuel was all stacked there in 44-gallon drums on a stone dock to the little boat haven where giant kelp waved its tentacles in the surging ocean.

A lone frigate bird was soaring over the haven, turning, and floating on the westerly, as though keeping guard over our fuel. I stood and watched the bird for a moment; and wondered whether they still flew over Cocos.

One of my men attracted my attention, pointed to the bird, and said, *"Manutara."*

The frigate bird—*manutara.* "The same bird." The Chilean Cat, and our aircraft; both at Easter Island. *Manutara* broken. I suddenly felt again the urgency for *Frigate Bird* to refuel and leave this island.

I turned back to the pile of drums and kept on pressing for their transport round to Ovahe Cove. They assured me the boats were on their way to Hanga Piko to load the fuel.

I arranged for Harry to stay and keep the pressure on. Then we drove back over the island for the hills above the aircraft.

The cloud was higher above the horizon: cumulus with pink creamy tops. The sting was going out of the westerly. A pastoral peace had settled over the island. The sun-warmed air was dreamy, and sheep grazed quietly on the hills. I wanted, somehow, to accept the hospitality offered to us. A great occasion had been arranged. It could not be ignored. I stood with Father Englert, whom I had met when we came ashore, and explained my dilemma. I looked out to the cloud. It seemed so ridiculous to be concerted about the weather when the day was so fine and perfect. I felt awkward, like some discordant being, out of tune with this perfection. Perhaps if I could be back in an hour I could make it before the threatening change reached the island. I turned with Father Englert and we walked back for the jeep.

We were all in and about to start away when I made the decision. I had instinctively sighted the jeep against the cloud bank to watch its movement. There was no doubt. It was coming in out of the south for the island; making up quickly, with the blue-gray of weather underneath.

"I'm sorry, Father," I said, "but I must go back to the aircraft; now. There's a weather change coming in with that cloud."

We stood for a moment in silence. Then he said, "We understand. Tell us if there is anything we can do for you."

His kind smile of assurance symbolized the warm hospital-

ity of the people on Isla de Pascua. I made my way down the
track to the rocks.

So far, this threatening weather change was just a suspicion in my mind. There was no line squall coming, or anything like that; no evidence of a violent change. There was
just the cumulus building up in the south, with persistent
movement towards the island. Had the surface of the sea on
the west side been feasible for landing and refueling, I would
have taken off now and put down off Hanga Piko; but the
seas were still coming in there as they were in the early
morning. There was no action I could take to improve the
situation of the aircraft, so I just went out from the rock
basin and we waited aboard for the boats with the fuel.

It was about three o'clock in the afternoon before they arrived and we had the hose pipe delivering fuel to the tanks.

By this time the wind had changed to a light but freshening southeast breeze, and the swell was increasing, and light
showers occasionally drifted over the island. *Frigate Bird* lay
now with her tail to the rocks, slacking and drawing the anchor line as she rose and fell on the seas. Only the good seamanship of the island men in the fuel boats, and the constant
vigilance of *Frigate Bird*'s crew, prevented the aircraft from
being damaged by violent contact in the disturbed sea. It was
very clear to me that we should soon have to be in the air, or
the surface would be too rough for take-off.

It seemed that even with the best efforts of everybody concerned the passage of the fuel to the tanks was frustratingly
slow. I stood on the wing, watching the weather, and checking with Blue on the number of gallons we had aboard.

Here, out of Easter Island for Quintero, we had to take on
almost full tanks. With two thousand miles of ocean between
us and the mainland of South America, and only the small
island of Juan Fernández near the route, I wanted plenty of
fuel to provide for unknown head winds which might otherwise run us short before we reached the land. We were taking
on full top tanks of 1430 gallons, and 120 in the hull tanks.
Allowing for climbing, and coping with weather, that would
give us twenty-six hours' endurance at effective power and
therefore an adequate margin for head winds. That could be
extended by using the most economical horsepower, but this

could not be relied upon, taking into account the possible need for the use of high power to deal with weather on the route.

I was tempted to leave the bottom tanks empty, to make a quicker getaway. We had often made the crossing over a similar distance from California to Honolulu on the top tanks only, but there the wind and weather information came in regularly from ships and aircraft on the route, and the forecasts could be relied upon. Here, as far as we knew, there wasn't a ship, and certainly not an aircraft, within thousands of miles; so forecasting had to be made on reports from Easter Island, Juan Fernández and the mainland.

As the afternoon passed the wind and sea increased, and the whole situation rapidly deteriorated from one in which a previously devised and logical plan could be carried out with reasonable certainty, to one where only a precarious take-off might possibly be snatched out of what would normally be regarded as impossible conditions.

The old whaleboats from which we were refueling slewed and swung in the seas which hit the hollow shell of the aircraft with an echoing clang.

About five o'clock Blue reported that we had just under 1600 gallons aboard. It was impossible to check the levels exactly in the bottom tanks because of the movement, but the top tanks were up to the filler caps and there was a margin in the bottom.

I called to the fuel boats to clear away from the aircraft, and went down to see the Administrator and others who were waiting in another boat. Too much was happening for any formal departure. We passed a hand to each other and I called out, "We are going out to try for a take-off, but it may not be possible. Can we have a boat stand by in case we have to return?"

They waved in acknowledgment and their boat lay off and moved away with the aircraft.

I went forward and up into the pilot's cockpit. Angus was in the bow, and ready to haul in the anchor. I called him through the roof hatch. There was too much action for the earphones, with their leads getting mixed up with the anchor line.

"We'll start one engine, Angus, and you can take in the slack as she comes up over the anchor. Then take a turn round the bollard and we'll break it out with the motor."

"Very good, Skipper. All ready in the bow."

Blue was on the intercom; reported engines ready.

Jack was behind us, in a tough situation. He just had to take it. I turned to Harry and we grinned at each other. I don't know what he was thinking.

"Not so good, Harry. But we'll go out and have a look at it. May be better off the land, in deep water."

He drew in his seat belt, good and tight; I gave mine an extra hitch and called up Blue for the starboard motor. I let it idle, with opposite rudder and aileron, as she came up over the anchor. It came out easily and I let her slop into the seas, idling, while Angus pulled it up and stowed it. As he closed the bow hatch we started the port engine and *Frigate Bird* began to pick her way out to open water.

The wind was from slightly east of south, blowing up the coast at about twenty knots. The main swell was coming in from southwest, crossed by local seas caused by the change of wind. Every time I tried to give her some engine the propellers picked up the broken water and threw it showering over the aircraft. As she nosed down into the troughs I had to shut down the motors to keep heavy water out of the propellers. The whole thing was fantastic and ridiculous, but I kept on out, looking for a slant where wind and swell might let her get up and going enough to take the rockets.

She was heavy, and sagging each time into the seas, but we managed the run-up in a better patch, sliding down the side of the swell and slewing almost beyond control against full rudder.

I took her out about half a mile from the land and turned upwind to face the south cape of the island. Conditions generally were no better than in by the anchorage. I tried a test run, up to about half throttle. She plunged into the sea, completely submerging her nose so that green water came over the windshield and I felt the shudder as the propellers took it, and I instantly shut off to avoid damage.

It was raining now; bleak and gray with approaching night. The island seemed distant, and unable to offer anything to the aircraft alone in the plunging, heaving threat of the ocean. The air and security were close overhead, inviting me to attempt a take-off whatever the circumstances. I drove her into a turn with the starboard engine, to go back farther down the island. It was becoming difficult to bring her round now as the float buried itself deep in the rising seas.

A mile off the far end of Ovahe Cove I let her come round to wind again and we sat there in the cockpit looking out over the broken sea.

Then I realized it was impossible. Whatever the temptation to escape to the air, I knew that disaster was certain if I attempted it. Immediately the power was applied the aircraft would bury the propellers to the hubs in green water. She would never, even with the rockets, reach a planing attitude for take-off; and if by some chance she did, she would dash herself to pieces against the wall of a combing sea. It was the very thing I had planned to avoid in my repeated check on the efficacy of the fuel arrangements for Easter Island. The people of the island had done everything. They just didn't have the equipment to cope with the situation. It was maddening, sitting there with the rockets and everything set to go.

"I'm going back in, Harry. We'd break everything if we tried to take off in this sea."

Obvious relief was written on Harry's and Jack's faces. I called Blue and Angus and told them of my decision.

It was too late to taxi round the island, even if we could have made it in the rough water. Close in off Hanga Piko would be more sheltered now, but darkness was approaching, and we needed daylight to make that passage and to go in close and anchor at the other end. The big swell was still from southwest; still booming in, I imagined, on the west side. There was nothing for it but to go back in to water shallow enough to anchor, and try to ride out the night off the lee shore of Ovahe Cove. This was a prospect which gave me no sense of security at all.

The boats were still there as *Frigate Bird* approached, and they made signs to us advising us to stop farther out than our previous position, in the spot I had seen marked with an anchor on the Admiralty chart. The superior local knowledge of the Easter Island boatmen was a distracting influence, but on

the way in I had decided to anchor on the sandy patch close
under the bombora. The whole situation was so perilous for
the aircraft that I sought some method, however drastic, to
avoid full exposure to the force of the sea. By anchoring
close in off the rocks, almost in the scend of the bombora, I
reckoned the aircraft would have a chance of lying in water
where the first onslaught of the rollers had been broken by the
bombora before they came under her hull. In the brilliant
sunshine of the morning, with the offshore wind, this had
been a sparkling, beautiful place. Now it was a dark and boiling caldron with white scum on the surface where the tow
was coming out from the rocks.

I explained my plan to Angus, upon whose smart seamanship in the bow final success depended.

"I'm going right in, Angus, just clear of the surf, before
the rocks. Then I'll let her come round upwind, and anchor
in that patch by the bombora. It looks pretty horrifying, but I
think we can let the bombora take it for us there. You can let
go the big anchor first, then bring her up on that. Then we'll
get one of the boats to take the cable anchor out and drop it
off our port bow so we can haul out on it if we get too close
to the bombora. Maybe we can think some more after that.
Are you all set?"

"All set, Skipper. I'll hook the anchor over the bow hatch
now, ready to let go."

"Good, Angus, it'll need to be quick. There isn't much
margin. Jack can give you a hand on the winch end of the
cable anchor."

Both anchors were in fact the same; but we referred to the
one with the chain and manila line as the "big anchor," and
the one with the stainless steel cable on the winch in the bow
as the "cable anchor."

I think Angus enjoyed this sort of thing. Once he got wet,
he really got going; and he surely was wet now. He stood on
the sponson, holding on to the edge of the bow hatch, taking
the seas as they came. She was shipping water through the
hatch, but we could get that out of her later. We had to be
ready now with the anchor, to let go exactly where she
needed it.

Frigate Bird slid warily on, the men in the boats calling, waving, and warning us of the danger. I couldn't explain to them now.

I took her on in with the following wind slightly on the port quarter, to avoid any possibility of an uncontrollable swing to starboard where the rocks lay. The bombora went by fifty yards out from the port wing, the seas running over the sinister rocks hidden a few feet below the surface. Just as the rollers started to lift, before the breaking surf, I drew off the power on the port motor, started the swing, then pulled off the starboard. *Frigate Bird* slowly wheeled round in the broken ocean, where the froth and confusion came back in savage defeat from its onslaught on the rocks.

"I'll have to cut her with the switches, Blue. There won't be time to wait for the fuel cutoff."

She lay there edging up slowly towards the streaks behind the running seas. A touch on the starboard to bring her out a bit. Off again. The float crept to within a few feet of the broken water. This would do her, when she came back on the slack of the line.

"Let go, Angus."

I cut the switches. She stopped in a few feet, slowly bore away and drifted back. Angus took the tension and worked the rope to bite the anchor into the bottom. She was coming back on it, hard. He whipped it fast on the bollard. Now it was the bite of the anchor, or the surf behind her tail.

"Stand by to start engines, Blue, in case she drags."

"Standing by."

The rope straightened out from the nose, tightened, and held her. *Frigate Bird* lay steady on her anchor. Waves smacked under her bow, and she rose and fell on the swell, but for the moment she was fairly comfortable. I felt edgy and alerted, watching the run of the break on the bombora, back to the surf behind us, and out to the open anchorage where the boats lay. We just might get away with it, if the anchors held.

Darkness was coming rapidly now. We signaled for one of the boats to come over. He came within a few yards, then we passed him a line and he hauled in our cable anchor, the one that had come away at Mangareva. I showed him where I wanted it dropped and in a few minutes we had our second anchor out. Now if we could get a really big one with heavy tackle from the island, that would be the answer.

More signaling and gestures and the small launch came in, with an English-speaking man aboard. I explained what we wanted and he called back that an anchor would be brought out to us.

Soon afterwards two of the boats left, and headed back for South Cape, apparently to return to shelter before darkness. The third boat anchored in a position not far from us. There was nothing more to be done now. I looked out to the darkening cliffs where a few islanders still stood watching against the skyline. Then I went below.

The impression inside was bad. Broken water was smacking under the bow with shocks that went right through the aircraft. As she fell into the trough of a sea and the bow met the oncoming swell, the tail went down and its flat bottom hit the water with a clang that resounded through the hollow shell of the aircraft. I wondered how much of this she could take without something going.

But these effects of the sea settled into a series of regular noises and movements and we soon began to accept them as normal. I went forward to see how the anchors were holding and found her hauling back steadily on the line as each sea passed under her. The rope was absorbing the stresses and there was no tendency for it to snatch at the aircraft. The cable to the other anchor was slack, as it was intended to be. There were signs that the aircraft was fairly well established in her position. I began to relax now and to think of ordinary things like food and rest.

Down in the hull Jack produced some canned food from the cupboard over the fuel tanks, and we had bananas, oranges, papaw and other fruit collected at Mangareva. The indispensable coffee was brewed on the little electric stove by securing the pot with wire, and we soon began to feel a temporary illusion of warmth and security. From inside the aircraft the clanging of seas on the hull smothered the sound of the big swells bursting on the rocks behind us, so for a while we lived like people in a flimsy air-raid shelter, hardly aware of the turmoil and menace above.

The food and coffee loosened us up a bit and for the first

time I could think of the island as a place we were visiting
with an aircraft. The long grassy plain where *Manutara* had
landed seemed like a natural airstrip, needing no formation
work, but only foundation and surface to take the big inter-
continental aircraft. It was reported that the Chilean Cat had
been damaged when attempting to take off from the sea.
How could it have been brought back to its position under
the trees? There was some mystery there. I thought of her
again as *Manutara*, with the broken wing. Then of the frigate
bird soaring over our fuel drums. The legend of the *manutara*
bird. The figure of Father Englert standing against the skyline as I went back down the grassy slope for the track to the
rocks.

The tail smacked down in a trough with a bang that shook
the aircraft. It snapped me back to the present. But it passed,
and we were just there again, in the hull of the Cat.

Jack had been ashore most of the day. He had ridden a
pony over the island to deliver the Easter Island mail to the
postmaster. The official mail, of course, passed through the
regular postal procedure; but we had a few souvenir enve-
lopes with the stamps of each place visited, and these were
becoming really colorful by the time we reached Easter Island. At Easter Island we had them canceled with the seal of
the Isla de Pascua.

An unexpected discovery here was a fellow Australian, Mr.
Jack Lord, with a full Australian setting transplanted to this
eastern Pacific Island. He managed a sheep station here with
35,000 crossbred sheep. I had met him ashore, but my preoccupation with weather, fuel and a quick survey of the pro-
spective airstrip site had left little opportunity for the interesting facts of life on the island.

Sleep began to catch up with us there in the hull, with the
food and the warmth and the sense of temporary security. It
was Saturday night, twelve days since we had left Sydney, in
some past age according to our present impressions. Because
we had only the absolute minimum crew to work the aircraft,
we had slept last at Mangareva on Thursday night. I arranged watches for the night to allow some possibility of
sleep, and went forward to my seat in the pilot's cockpit for

this purpose. There I could be in action immediately for any emergency.

But I found that the freshness of the wind, the clanging of the waves on the hull, and some fundamental need to penetrate the darkness for vision of the bombora, the island whaleboat, and the land, kept me awake and peering into the night.

The whaleboat had gone.

This seemed ominous; but it wasn't long before he returned and hailed us with news that he had the anchor. My instant reaction was to stop him attempting to come alongside. From memory, I imagine the boat was about twenty feet, heavily built, and difficult to maneuver.

I soon saw the helmsman was ahead of me. The boat rounded up to the wind well off the bow and I could see them preparing to pass us a line. We took this over the bow, hauled it in, and made fast temporarily over the bollard. There was no need to direct the whaleboat men. They moved out and I saw them pass the anchor over the side. A rain squall came down on us and the boat disappeared in the darkness.

A few minutes later I saw them, anchored again in their previous position. These men were going to stand by us through the night at the risk of their lives. We took the line from this third anchor and secured it to the cleat on the sponson. So we had now three separate anchor systems, each secured to a different attachment in the bow of the aircraft; the manila to the main bollard, the cable to the mooring pennant, and this line to the sponson. If one carried away at the aircraft, the others wouldn't go with it.

I went back now to try for some sleep on my bunk in the blister compartment; but it didn't work out. Each time the tail came down on a sea she shuddered badly, and I imagined she might start to leak at the tunnel hatch, which was taking a terrific punching. I left open the bulkhead door to the tunnel compartment and kept this situation under observation. If she collected real water in this compartment she would start to go down by the tail, so it was necessary to be alert and ready with the pump. I just sat there on the bunk, numbed with waves of sleep and jerked awake by the critical needs of the aircraft.

Out under the starboard wing seas drove past the float where foam from the bombora lightened the night with white streamers. I looked up to the sky for signs of a break and the moon, but there was nothing.

At regular intervals a long-drawn, eerie cry went out from the whaleboat, like the wail of something lost in the sea. It passed away on the wind, and was lost itself in the boom and crashing noise of the surf on the rocks astern. The cliffs were a dark shadow in the night behind us.

Each time this cry went out, it seemed to stress the precariousness of our situation and to leave the aircraft with a greater sense of exposure. The wind was gradually increasing in force and the sea making up.

"Ai-y-e-e-e-e-e-e—"

The cry went out to the cliffs and away to nothing. It must have been a prearranged signal to the land, that the aircraft and the whaleboat were still there on their anchors.

I will not dwell on all the detail of that night. It was the worst I have known on the sea or in the air. I doubt if any aircraft has ever survived such conditions, at anchor in the open ocean.* All that could be done for survival had been done. We could only watch and wait for some sudden emergency.

Dawn came with the wind more in the east, and squalls of rain. The cliffs behind us crept up out of the darkness and showed the aircraft to be perilously close to the surf. She had dragged the anchors in the night, which I had suspected from the position of the whaleboat. But then the boat seemed to have passed from her position more astern and it was evident that she too had dragged her anchor. So it had all added up to nothing, and we just lay there in the fading darkness, listening, watching, and waiting.

Soon after dawn I saw some action in the whaleboat. They were hauling in on the anchor. I could just hear the thump of her heavy-duty engine, getting under way. It looked as

---

* The first *Frigate Bird* and the base Catalina 603 had, of course, survived the hurricane in the lagoon at Clipperton Island.

though she was leaving us; but she fell away on the wind and headed slowly in, shooting the seas, for the sheltered corner behind the rocks. There must have been a tremendous surge there, but she went on in and disappeared round the corner.

In a few minutes she headed out again for the aircraft. Something was happening. Perhaps she was bringing us some heavy gear.

The boat came up to us and stood off beyond the wingtip. The helmsman held up something in his hand and indicated that he wanted to pass it to us. I couldn't have the boat alongside. He understood my signals, let down a shade on the wind and worked the boat in behind the wingtip. One of the boatmen threw something and it landed through the open blister. I picked it up; a piece of paper rolled round a small lump of lead attached to a fishing line—obviously a message. I waved a hand to the boat and signaled them away from the aircraft. Then I undid the line, opened up the paper and spread it out to read.

It was a message from the Administrator of Easter Island; and it read:

> Message for Capt. P. G. Taylor.
> Until midnight tonight the following people are waiting to speak to you on the wireless; Director of Aeronautics General Gana, and two members of the British Embassy.
> They are going to Quinteros Air Base tomorrow to await your arrival and wish to inform you that the best hour for arriving in Quinteros would be at 1300 hours continent time.

A big sea came under *Frigate Bird*, went surging past, and the tail came down with a crash that exploded the utter hopelessness of this message. There, in Chile, they were waiting for us; confident, even suggesting the best time to arrive. Here at the island I was beginning now to consider the advisability of abandoning the aircraft, since it was unlikely that anybody would walk away from the wreckage if the anchors went. I read on.

> For myself, as Chief of the Air Protection Service, I would like to salute you sincerely and congratulate you on your splendid flight and wishing you all the best of luck on your last hop to the Continent.

Tomorrow at 9 A.M. Continental time, the Naval wireless station at Valparaiso will send you the weather forecast.
Signed Cmdte Swester.

2° Message.
Cmdte Parragué sends his best wishes for your trip and is awaiting you at Quinteros.

I don't think I have ever been nearer blank and hopeless despair than I was at the moment when I finished reading that message. I was beyond hostility about the failure of the fuel arrangements; and that is bad, when you can't feel hostile any more.

Parragué would know, if he could see us. Parragué, whose *Manutara* lay wounded on the island while Chile was waiting to receive him back with honor from his flight. Now it seemed that the critics in Australia had justified themselves; but in what maddening circumstances, when we had made proper provision for the risks we knew existed at the island, had properly dealt with the conditions when we arrived, and had put the aircraft in a safe situation for refueling and take-off. Explanation of the cause would be futile. If *Frigate Bird* were lost, we had failed; and that was about all that would be known of it. I began again to be capable of hostility. The dreary fatigue of hopelessness following two days and nights without sleep left me, and I began to press my resources for some way out. There had to be one, somehow.

I went up on top of the wing, and stood with the wind in my face. It was due east now, moving through from the original southerly change of the afternoon. If we could hold on till the cycle moved round through north and the wind came off the land again, the seas would flatten down and we might be able to take off as we had originally intended on the morning of arrival.

I went down to the bow to check the lines to the anchors.
The manila was taut and holding; but both others were slack. They had been tight before when I checked them in the torchlight after she had dragged from a line on the whaleboat. I bent down to take the strain on the rope to the

island anchor. It came away loosely in my hands. It was gone.

So was the steel cable to the Catalina anchor.

Rope and cable were tangled under the surface below the bow, but the anchors were gone. Only the manila was holding her, and each time a sea came under her bow and drew her back it stretched and tightened, quivering with stress like the line from a rod to a fighting fish.

The perspex was broken in the bow turret where the manila had squeezed down on the bollard, and a hole was left where every passing sea threw water into the aircraft. I looked inside and found water in the bow compartment up to the catwalk. Things were really beginning to catch up with us. Something would have to be done; and quickly.

A really heavy anchor and a chain. That was the first.

Harry. Yes, Harry had better go for that, and Angus with him.

I went back down, got the two together and gave them the score.

"As things stand at the moment the aircraft's had it, unless we do something very quickly. Two anchors have gone, and the last is holding, precariously. Water is coming in the bow. We either have to abandon her, now, to keep ourselves out of the surf on the rocks, or secure her till we get an offshore wind change for take-off.

"I think the big line will hold her for a while, and it's worth having a stab at getting a really good anchor, and heavy chain. Best you go after that, Harry; and you, Angus, make sure it's what you want in the bow. Get it, somehow, as quickly as you can."

I could see that Harry didn't want to leave us with the aircraft. He was silent. But he knew it was best, and he went. We hailed the whaleboat and he lifted his anchor and got under way.

This was the tricky part of it, coming near enough for Harry and Angus to jump; and not bashing in the hull. It was a risk which now had to be accepted.

The whaleboat came round astern, and crept up very slowly just clear of the tail plane, which was coming down close to the seas. It was a perfect piece of seamanship. He sneaked the boat in behind the wing, keeping the stabbing bow clear of its thin covering, and edged in for Harry and Angus to jump.

They made it, from the blister.

He let the nose come away, and before the boat could be thrown back under the tail he caught her with the engine and drove her out in the clear. We had gotten away with it.

Blue and Jack and I watched the boat on her way from behind the rocks. She rolled and slid down the seas, disappearing in the troughs, riding high on the crests again, till she passed from sight. Then we took stock of our own situation.

Blue and Jack got on to the pump in the bow and started to suck out the water. I managed to ease off the rock-hard line on the bollard, and let out some more slack for the anchor. The manila was stretched to about half its original thickness now; but it looked strong still and able to hold against the seas. I thought of the amusement this heavy gear had caused at home when we loaded it into the bow. Now I was inclined to wish I had brought 4-inch line instead of 3½. I just didn't think too much about the other end, but I imagined the chain would be taking the chafing on the rocks where the anchor had probably fetched up after pulling through the sand.

Back on my bunk in the blister I had a small cushion from the cabin of our boat at home. I took this forward and stuffed it in the hole in the perspex. I didn't know then that this cushion was going to save the aircraft within the next few hours.

She had taken no water of any consequence except in the bow, and Blue and Jack soon had this dry. The cushion held, and prevented the seas breaking in through the hole. Things were under control again; but it was really blowing, and showing no signs of improvement except for the very gradual swing of the wind. We just had to hope the anchor would hold till Harry and Angus could get some heavy gear.

The last resort was the engines, so we went to our stations for action in the emergency of the last line going; Blue to the engineer's tower in the center section, and Jack up front with me in Harry's seat.

I knew it couldn't last; just sat there balanced on a knife

edge of anticipation. The surf was roaring over the bombora now, rushing by and burying the starboard float. It was a savage, sinister thing, that bombora; but it had done well for us, breaking the worst of the seas as we lay precariously between it and the surf behind.

We watched the anchor line taking her each time the nose rose on a sea. They were breaking now, and coming in steeper. Each time a big one came under her the rope, loosened momentarily by the plunge in the troughs, would flick tight and quiver, rigid as a ramrod as it took the strain of sixteen tons of aircraft thrust back by the sea and the drag of the wind. I wondered which would go—the line, the bollard, or something at the anchor end. I looked out to the whaleboat, which had returned. Light, and with little resistance to wind or sea, she was riding it out well, in worse conditions out at the anchorage. Breaking combers came under her; her bow rode high in the air, shedding the ocean from her sides; then she would go down, disappearing in the trough. A moment of suspense, then she would climb in sight again.

Ahead of us a big swell was approaching, one of those freak seas that came up out of nowhere. Fascinated, we saw it rolling towards us, breaking at the top in a white wall of foam that roared down for the aircraft.

This was it, all right.

We sat tensed, suspended in a state of futile inaction, brought up and stopped like characters in a movie that is suddenly a still.

Out of the chaos that broke over the bow I felt the aircraft rise and myself almost lying on my back in the seat. She was lifted and thrown back like a bird on the wind, and as the foam passed over the cabin my eyes went straight for the anchor line. It was still there; but the rope was slack and the aircraft was sliding back, not stopping now as the sea passed under and away for the rocks behind.

"Engines, Blue," I shouted, "starboard first. The anchor's gone."

Over our heads the gray blades of the starboard propeller moved and slowly began to turn; the engine awakened instantly to sudden urgent life, and swung into action to save the aircraft.

The port engine came in. I took the throttles; felt the movement of the unlocked controls. I didn't know where we were going, but we'd know in a minute whether the engines would take us out of here. Hurrying with power in the wing above us, they checked the abandoned, hopeless drift for the surf, and brought us up facing the seas, but still moving back. I pressed on some power to beat the seas and we faced up to it, thrashing and blasting the ocean with broken water from the propellers.

Each time a sea came in, the propellers bit into it and shook the aircraft with vibration. It was impossible to use the power. Green water as hard as rock to the spinning propellers would have torn the engines from their frames.

By using the throttles with each opportunity when the propellers were clear of the water, I found that she was just beating the forces that sought to drive her back for the rocks. That was fundamental. We were going somewhere, not just out of control, blowing away to destruction.

She crept out past the bombora, slowly but convincingly making headway for the sea. I felt the cold challenge of water and the wet taste of salt in my mouth as the spray found its way in through leaks in the screen and the cabin roof. Below us, water was swishing again in the bow compartment, rising above the catwalk as each sea that came over the nose found its way in through the forward turret.

"Better get some of this water out, Jack. See if Blue can leave the engines, and get busy with the pump. We'll have to keep the water out of the bow."

Jack was away below in a moment; relieved, I think, to have something to do.

At first they seemed to be making no impression on the water, with the emergency hand pump that had to be used now because the main bilge pump had failed. I watched the sea, the propellers, and down into the bow compartment where Blue and Jack sweated, wet and steaming, at the pump.

Behind us, the land was very slowly moving away and there was space now round the aircraft. The immediate danger, of destruction on the rocks at the base of the cliffs at Ovahe Cove, had passed, and I began to wonder what could be the next move. The water in the bow was the deciding factor. I was relieved to see that they were now slowly gaining on it with the pump. Clear, for the moment, of the coast, I kept the engines idling, just holding position so that the bow rode over the seas.

From a hopeless position we were now at least a partially going concern again; but ahead of us were two thousand miles of ocean and around us a sea in which it was impossible to taxi across wind to make a passage round for the lee of the island. It was a vastly improved but still embarrassing situation, with an aircraft on the ocean, the engines running, and no place to go.

The wind was blowing directly onto the land, so we could not ride for long on a sea anchor even if it were possible to rig one with drogues. There was no alternative but to keep off the land with the engines, and we could hardly expect to do that indefinitely till the weather changed.

I sat there, feeling totally ineffective; trying to come up with some sort of idea which might lead to an effective move.

It came out of desperation, but with a slightly humorous twist that lightened my mind with an objective and the prospect of some action for it.

"I'm going to have a stab at sailing her, Blue. It's impossible to taxi to the end of the island because we can't turn crosswind in this sea; but there's a chance that we could tack up the coast by using one engine to work her out; then cutting it and sailing back with the reversed controls. We'll take her out a bit farther; then try to sail her backwards. If you cut the starboard motor now I'll crab her out on the port."

Using throttle on the port engine, right rudder, and right aileron down to cause as much drag as possible on the right wingtip, and with the starboard engine stopped, she began to work her way out to sea.

Every time there was a lull in the wind and sea that showed a sign of being able to turn the aircraft, I gave her a burst on the motor and she swung a little to starboard. We watched the land behind us, and marks there did seem to be moving slightly across our track. It seemed to be working on the outward passage.

Half a mile off the land we cut the port motor and started

to sail her backward. With left aileron down, and full left
rudder, she slewed across the seas and sagged backwards.
Then a sea got her and pressed the port float under to the
struts. I reversed the controls and the wind brought her back
level and out of that one. I soon found a way to anticipate
this dangerous effect and catch her so that she took the seas
and sailed back fairly level. It was precarious, but it was
working, and *Frigate Bird* was perceptibly gaining ground in
the direction of South Cape. In calm water there is no difficulty in sailing a Catalina backwards, but when steep and
heavy seas are added to the wind it is different, and each
backward surge is filled with suspense, like a surf boat riding
the breakers for the beach.

When she had sailed back near enough to the rocks, Blue
started the port engine again and we set out on the plunging,
swaying passage seaward. I tried to taxi across sea in the better patches, but each time I attempted to turn her the lee
float plunged under to the wingtip, and the propeller blasted
the air with green water, shaking the aircraft with the threat
of something breaking. After several attempts to make good
some distance in this more direct way I abandoned that
method for the tack out to sea with the engine, and in again,
sailing for the land. On each board we made only a barely
discernible distance along the coast, and with three miles to
go to round the end of the island, the prospect of making it
within the day did not look good. But the temptation to beat
her crosswind with the engine had to be resisted, because always in the background now was the distant hope that she
would fly again, and it was not merely a matter of saving the
aircraft from the ocean, but of considering the stresses imposed in her structure on this strange passage for the shelter
of the island.

At the end of the third stretch out to sea, when the method
of working the aircraft was showing some results, I began to
wonder how we might make better progress and avoid the
danger of being caught still at sea on the exposed side of the
island at the end of the day.

Down over the bow, the lines from the anchors were still
hanging in a tangle of rope and cable that interfered with
maneuverability, and retarded the movement of the aircraft

sailing backwards. With the engines stopped, and *Frigate Bird*
riding on the swells, I went down over the bow, and managed
to get a fairly good position there on the sponson. The seas
were washing over, but the water was warm and refreshing,
and I was able, holding with one hand, to free the lines with
the other.

I managed to clear the manila from the cable, and haul it
in through the bow hatch. It came right through to the end,
still carrying the shackle which had secured it to the chain.
The heavy shackle pin was gone; bent, then sheared at the
head. The anchor and chain lay shackled together on the bottom at Ovahe Cove.

I went down again to untangle the other line and the steel
cable, and clear them from the bow of the aircraft. I couldn't
do it with one hand; so I was working at it with both and
balancing on the sponson when a big sea came in, surged
over the bow, and just lifted me away to the ocean. I found
myself washing away down the side of the aircraft, managed
to get a foot against the hull and push away clear of the
pounding tail.

When the first surprise passed I wasn't much concerned
about being in the sea. I had on only a pair of shorts, I
couldn't have been wetter than we all were up on the bow,
and the ocean water was warm but invigorating. Strangely
enough, I didn't think about the sharks which would certainly
have been hunting in numbers off the coast of Easter Island.
The indifference of the brown-skinned amphibious islanders
to these monsters would not normally have reassured me, but
we had so been reduced to the fundamentals of survival for
the aircraft, and the concentrated action to be taken for that
single-minded purpose, that it simply did not occur to me to
be concerned about my own situation which, in normal circumstances, would have horrified me.

I think this incident had some dramatic publicity afterwards, but it was more humorous than anything, except for
the problem of getting back aboard the aircraft, which was
considerable. The ladder from the blister compartment would
almost certainly have rammed me very effectively in the
ocean as the soaring, plunging tail came down on the sea; the
chine of the hull was coming out of the water and smacking

down in a way that would have flattened me if I made the mistake of being sucked under it when trying to get aboard. The sides were smooth and there was nothing effective to grab as a means of swinging up, and back on the aircraft. I saw a look of urgent horror on Jack's face as he looked out and back as I finished somewhere down off the tail.

Back to fundamentals again, with all the trimmings gone, I swam away clear of the aircraft first, and then started to think about how to get back. There was a rush of activity aboard. Jack disappeared below and came back in a minute with a piece of spare line which he threw over the side. It was about the best thing he could do. I called out to him to make it fast well inside the aircraft so I would not drop back in the sea. Then I watched the movement for a while to see just how the seas were running under the chine. There was a space when it was well submerged, so I swam in, waited for the right moment, and managed to haul up and onto the sponson.

Another situation had passed. Jack told me afterwards that he had said to Blue, "We've got to get him back on the aircraft. What's going to happen if we're left out here on the ocean?"

About this time Harry and Angus, having located an anchor and chain, had returned to Ovahe and found the aircraft gone. The whaleboat also had disappeared, though we ourselves had seen it go back for shelter round the island.

They both thought we had been wrecked and lost in the caldron of ocean below the cliffs, till looking out to sea they sighted the incredible spectacle of a Catalina beating out from the island with the rollers bursting on her bow and coming back over the aircraft in showers of spray. They saw us plunge into the big troughs and disappear, apparently swallowed by the hungry ocean; only to rise again, showing the silver wing against the gray of the broken, shadowed surface.

Harry and Angus realized what we were trying to do. They made their way back over to Hanga Piko and arranged for the anchor and another refueling on that side of the island.

Conditions began to improve slightly; or perhaps breaks in the cloud, with sunlight on the water, only made the whole situation appear less grim and menacing. But our progress,

though definite, was slow and still left doubt about our reaching shelter before the night. So I decided to jettison a large
quantity of fuel and attempt to force a passage across wind
with the port engine. Again, it was a question of choosing the
lesser evil. I thought it better to force the aircraft, lightened
to some extent by jettisoning the fuel, than to be left in an
exposed position when darkness came over the sea. I discussed it with Blue as we drifted back on one of the sailing
tacks, and we agreed that the risk of fire from fumes after
jettisoning on the water, and of the dump valves failing to
seal again when we closed them, would have to be accepted
for the advantage of lightening the aircraft. Blue pulled the
dump valves, and we let go five hundred gallons of clear,
blue-green 100-octane gasoline.

To reduce the risk of fire, we let her sail on back as far as
possible before starting the engine. Then Blue engaged the
starter. We held on for a moment, though the sailing tack
had taken her well across wind from the fumes of the jettisoned fuel. She picked up and I took her away, still moving
across, on the engine.

The advantage in maneuverability was evident; and this,
with the conditions which now were noticeably improving,
enabled me to blast her round broadside with the engine,
make a rush along a chosen swell, and gain some distance before I had to shut off again and let her come up to the wind.

We managed in this fashion, combined with the regular
sailing and engine tacks, to make a windward position off the
end of the island early in the afternoon. Blue and Jack were
almost continuously on the pump and the bailer, because we
were unable to prevent the breaking seas from forcing water
through the bow turret. It was a laborious process; but it had
to be done to keep the aircraft afloat. The cushion I had
stuffed in the hole in the perspex was still wedged tightly in
and holding.

There was a magnificent sense of triumph as we cut the
port engine at the end of the last upwind tack along the island. I looked downwind and judged that on the next sailing
tack she would run through the passage between the needle
rock of Motu Kao-Kao and the high cliffs under the crater of
Rano Kao. The seas were breaking heavily across the greater

part of this passage, but I could see an unbroken run close up to Motu Kao-Kao. If we could make this passage and sail through it, the last hazard would be passed. Beyond this pinnacle of rock and the white mist of the broken rollers that lay between it and Easter Island was the open sea on the sheltered side of the land.

I was beginning to build up the structure of things again in my mind now, and it seemed not beyond possibility that we could refuel again and take off before darkness. My urge to this objective was strengthened by signs that the wind was moving to work through its cycle. Though the west side would now be sheltered and, I imagined, possible for take-off, by tomorrow or even through the night it might again be exposed to wind with a westerly component.

As we sailed back on this long, downwind tack for Motu Kao-Kao we ate most of the fresh food within reach in the aircraft. Jack had acquired some roast chicken for our supper last night and with it some salad and fresh fruit. With the engine stopped and the aircraft surging backwards quietly on smoother seas, we all sat up on the bow and ate last night's supper. Now, in the brilliant sunshine, warm on a sparkling sea, no more water coming into the bow, and the wind setting her well for Motu Kao-Kao, we really enjoyed that food. I trimmed the rudder control and sailed her on the ailerons, sitting in the sun above the cockpit with my toe on the wheel through the roof hatch.

But as Motu Kao-Kao came in there seemed little space for the passage of the aircraft and I wondered whether I hadn't chosen with too fine limits. Beyond the island a huge, long swell, born in some distant disturbance, was still coming in from southwest. Riding this were the more local seas of the easterly wind. The result was a roaring surf breaking from the foot of the island cliffs right across the passage to within a short distance of Motu Kao-Kao. It was beginning to look as though we'd have to pick her up with the engines and lose vital time making several more tacks to clear Motu Nui. But there was deep water right up to the pinnacle of rock whose base we were aiming to clear. We could hold on and if necessary pull out at the last moment with the engines. Also, there was good sailing control now that she had stopped the headlong slides down and across the rougher seas; so we could sail to finer limits than before. We kept her going, Blue back at the engineer's station, Jack and I sitting up in the sun on top, watching and judging the passage of the starboard float by the rocks of Motu Kao-Kao. Beyond the port wing the big rollers were lumping up, sometimes uncertain of their next action, but some rising suddenly, crumbling at the top, and breaking in a roar of foaming surf.

I could see she was going to make it, close by the rock.

She slid on. There was no doubt about it now. The base of the pinnacle was drifting by the wingtip float. I looked on up to the peak with a general interest. Over it a bird was soaring on the wind. Without looking down, I touched Jack for his attention and we both looked up to the whitened rock of the pinnacle drifting past against the sky, and to the lone bird riding on the wind.

"We're right now, Jack. You see the bird?"

"Frigate bird." He recognized the swept-back wing.

"Yes, Jack. Always the *Frigate Bird*."

We sailed away southwest of the island, headed now tail first for five thousand miles of ocean. It was a strange feeling with the wind blowing us away from the land. I let her go a mile beyond Motu Kao-Kao, then we started the starboard engine and tacked the other way, to port.

The easterly was whipping round the island as we came in close under the cliffs; still too much sea and wind to taxi directly up the coast with both engines. We made two more tacks, with the starboard engine, and finally reached the shelter of the land. The big southwest swell was bursting on the rocks off Hanga Piko, but the surface was smooth, and I could see us getting off with the rockets. We started the other engine and taxied smoothly up the coast close in under the cliffs that gave us complete protection from the east.

About two miles up the west coast from the end of the island I could see a niche with perfectly smooth water and the point on the chart shown as Pointa Baguedano. I took her in for this spot. We could refuel here, and be away before sunset.

As we approached the bight in the coast I could see a boat there. It was the black whaleboat, standing by to receive us with an anchor. We needed it, because we had no anchor now and just kept taxiing till we were near the boat. Then we cut the motors and hailed them.

They came over with a line and we were soon hauled up on the anchor, and secure. It was four o'clock in the afternoon, two hours before darkness. We'd have to get moving with that fuel.

I called Blue and we stood out on the cabin top. To go on immediately was outrageous, but I knew the others would be with me, whatever I asked of them.

Blue gave a cagy, inquiring look. He had something on his mind. "Look, Skipper," he said, "I heard some very strange noises in that center section when she was straining in those seas coming round the island. I ought to open it up and have a look inside before we fly her again."

I knew he was right. The center section, the wing strut attachments, the floats, and the retracting gear: a dozen things. All should be inspected. But I also knew in my heart that the weather would be following us right around; that if we stopped to make a thorough inspection we could not get away before sunset, and a night take-off without clear vision would not be feasible from the open ocean. I knew for certain that if we stayed here on the anchor we'd lose the aircraft. We just had to get into the air, *now*, and fly for South America.

"I know, Blue. You're right. But we have to take off within two hours, before sunset. If we don't, we've had it; and we'll lose the aircraft."

"Very good, Skipper. I'll get started on the fuel."

Blue was silent then; but I knew what he was thinking. From a pilot who wouldn't take the slightest avoidable risk on the airlines, I had suddenly appeared as a reckless individual deciding to take off without an inspection in an aircraft that had been severely stressed in a way for which it was never intended. I knew he understood the reason; but his professional instinct was outraged and his acceptance was expressionless.

Harry and Angus had the fuel drums in boats. It was soon alongside, and Blue had it going up to the tanks.

I went down and laid down the track chart for Easter Island–Valparaiso, checked over the chronometer, sextant, and my books and tables. We'd lay a course for Juan Fernández, 105 degrees True from Easter Island. I corrected the course for variation and deviation, guessed the drift I thought we'd have in the beginning out of the island, made the allowance, and clamped the course on the master compass on the navigation table. Then I lashed everything down for the take-off, and went up top to see how things were going.

Two hundred gallons still to go in at five o'clock. That looked all right, as though we'd make it. I got Jack to dig out the last of the cigarettes and pass them to the men in the boats. They were all smiling and friendly and glad we had survived the night and the passage round from Ovahe Cove. But on the cliffs people stood against the skyline. They were still; awed; and watching; for it was known now that *Frigate Bird* was *Manutara*.

As the sun was setting at the Isla de Pascua and still, in my vision, high over the eastern beaches and the bush of Australia, the main tanks were full again. We passed the anchor line to the whaleboat and let *Frigate Bird* drift back on the wind off the land. When she was clear of the boat we started the engines and turned her out to sea.

I was standing up through the roof hatch when a puff of wind took my cap and it blew back through the propeller. I had to recover that cap. It was old and battered now, but it carried the badge we had designed for the flight: the red waratah of Australia, with the seabird over crossed leaves of the great spotted gums that reach up to freedom in the skies.

I called to Blue to stop the engines. The cap was still floating behind the aircraft. One of the boats went out and picked it up. It had been cut to shreds by the propeller, but the badge was still intact. I laid the remains of the cap by the navigation table and went back up to the pilot's seat. In a few minutes we had the engines running again. I headed her out for the open ocean and began to fit the surface conditions into a plan for take-off.

The wind was directly off the land. The swell, huge and long from the southwest, was running across the wind. Under the land the conditions were excellent. I considered a crosswind take-off close in under the cliffs along the coast; but discarded this idea because the direction was across the swells, and sudden, vicious black puffs were whipping down from the cliffs and hitting the water, crosswind, in the area we would have to use for take-off. It was best to go out to sea, far enough off the coast to clear the cliffs on a take-off along the swells towards the land.

As we taxied out into the sunset the surface began to lift in a roughening sea where the wind again had a free run out from over the land. The easterly swell also was reaching through from the southern end of the island and crossing the big southwest roll that came in from the heart of the ocean.

There was no way to find good conditions for take-off within reach of the aircraft at the island. I would have to judge the minimum distance in which we could take off towards the land, and clear the cliffs and the island. As we went on out it became obvious to me that this margin would have to be fine, because the surface was rapidly deteriorating as our distance from the land increased; and out against the close horizon we could see the seas lumping up in conditions little better than those from which we had escaped at Ovahe Cove. There was only a narrow stretch of sea, along this west coast, where take-off would be possible.

After the run-up and final check of the aircraft I looked in towards the coast, measuring the take-off in my mind.

There was a good breeze. That would help to lift her off the water. The rockets would shorten the take-off, which otherwise, with the overloaded aircraft, would be more than a mile. But we had to clear the cliffs. Suppose one or more of the rockets failed? They had been immersed in seas all through the night and the day, till we rounded the island and they dried off. Harry had checked all the leads and connections. I would have to assume they were going to work and give the aircraft the added thrust she would need to clear the cliffs. If they didn't, could I turn her away below the cliffs?

No. That was obvious. If I took her out far enough to allow for rocket failure, what then?

I looked out to the leaping surface, not far beyond the aircraft now. No. The distance would have to be judged for the minimum run that would clear the cliffs on a rocket-assisted take-off. There was no compromise.

Already the slap of the seas was clanging under the bow. I looked back to the island, seeking a better way out where I knew there was none. The cliffs were there, unchanging, and clear-cut in the evening light. With the idling motors waiting, I measured up the distance in my mind. She'd do it if the rockets worked; but not from any closer.

"Right, Harry. All set to go?"

We checked across the cockpit.

"Engines ready, Blue?"

"All ready, Skipper."

"Right, Blue. We're going now. Wait till I let you know before you jettison the rockets."

I glanced back to Jack and Angus; gave them the thumbs-up signal.

"Harry, I won't know how I'm going to use the rockets till she's in the take-off. I'll let her have them as she needs them. Wait for the second selector switch till after you see me press the button for the first rockets. Then select the second so I can use them when I want them. O.K.?"

"All O.K." Harry moved to get comfortable in his seat, and waited.

"All right, Harry: give her the power; the full treatment* if she needs it."

Harry's hand went up to the throttles and the engines rose up and flung their challenge at the cliffs as he pressed the levers forward.

I took the controls to my hands and feet and joined with the aircraft, ready to supply her needs for flight. The sea came over the nose and I held her on to the gyro heading till it cleared and I could see the island ahead. A sudden thought of how near it was. No good. Forget it. We're in the take-off.

The nose was up, riding high out of the water now. A sea was coming at her.

Rockets!

I thumbed the button on the control column. She surged ahead and flattened down, charging to meet the sea. I lifted

---

* Above normal maximum take-off power.

her a shade to meet it. There was a surge. The engines were there roaring, and she was away fast on the surface of a big swell that was passing under her. I watched the airspeed indicator. It was registering now, gaining speed. Another cross sea was coming at her; a big one. It hit my mind like solid water. She had to fly now. I held my thumb above the rocket button, waited a moment till the sea was coming in for the nose. I knew it would bounce her. She'd have to stay in the air or the bottom would come out when she hit again. I pressed hard down on the button and held her tight. The rockets hit her; and then the sea. With all rockets going and full power, she hit, bounced, and lay on the air. I seemed to hold her in flight through my hands on the control wheel. She sagged and my mind and hands seemed to support her, trying to keep her off the sea without stalling. She had to keep flying now. To touch the sea again would be disaster. But she held on the air with the power and the rockets, and I could feel that the wing was beating the weight.

As the rockets expended the last of their power she had the speed and began to fly. Secure in the air, I held her down for more speed, still holding full take-off power; and as the cliffs came in I knew she had won.

"Climb power, Harry," I shouted across the cockpit, and drew her up on a steady climb to pass over the island.

The rockets had worked. Harry's meticulous work, and Angus's and Blue's, had kept them serviceable and ready to go, through everything. As *Frigate Bird* roared over the cliff-top where the watchers gazed up in wonder to the aircraft that had risen out of the ocean, I looked down and back to the cloud of rocket smoke that was drifting out to sea on the wind.

The grassy hills of Easter Island passed under her, and a wild exhilaration lifted me to freedom. Harry and I looked across the cockpit. He reduced power to ease the engines for a long and steady climb. I turned to see Angus and Jack below and we all looked what we felt.

The cliffs of Ovahe Cove passed under the starboard wing and we flew over the gray and broken sea. It was a good place to jettison the rockets.

I called up Blue. "You can let go the rockets now, Blue. It's a good place to dump them."

"Couldn't be better, Skipper. Rockets gone."

I didn't see them go, from the cockpit; but Jack, from the blister, saw the big metal cylinders leave the aircraft and fall, effectively bombing the sea off Ovahe Cove.

As *Frigate Bird* climbed away into the east, I handed over to Harry and went back to the blister compartment to see the last of the island.

There was still color in the western sky, but the shadow of approaching night was on the ocean. Already the Isla de Pascua was fading into the gray mist below the scattered cloud. I looked out to the colored west and experienced a calm feeling of thankfulness for the escape we had had from the sea; for the smooth security of the evening air where the aircraft was set in her orbit again, and all the scattered ends of our life-system had joined again in the harmony of united purpose. The sound, the touch, the sight of the aircraft, ourselves, and all the space around us, spun with the endless rhythm of that other sense that combined it all in a single consciousness of eternal life and purpose.

"At the going down of the sun, and in the morning, we shall remember them."

I looked back through the transparent blister, into the red light that shone with life above the horizon, and knew, as I often have in the air at the rising and setting of the sun, the presence of those who had passed from Earth to this fulfilling revelation that comes in the drumming of the motors and the lights of space.

I looked back into the gray depths again. The island had gone. There was only the dim, dark surface of the ocean and the white islands of cloud that drifted by a thousand feet below us.

Relaxed and ready for the night, I went forward into the hull, where Jack had food heating on the electric stove. I could see Blue's feet on the rest above, where he was sitting at the engineer's station. Angus was crouched at his radio, listening and stubbing a cigarette on his jam-tin ash holder.

Harry's hand came round through the bulkhead door, signaling that he wanted to say something. I leaned through and found he wanted to know about the height.

"The tops look higher ahead and rising into the distance. I've eased her up a bit but it looks as though we'd have to go to ten thousand to clear them."

"I think we'll hold on as we are, Harry. She's pretty heavy yet to push up to ten. If it gets too rough we can go down and stay under it till we see how it works out. I'm not much concerned about a position for the first few hours. We can take some stars later when it clears, and see where she's fetched up."

We stayed at six-five and soon were brushing through the tops of the cumulus.

In ten minutes we were immersed in cloud, passing occasionally through a gap where a black canyon of night went down to the ocean. Sometimes a star appeared overhead, flickered in a light edge of mist, and was wiped out again as we entered the close darkness of cloud.

The air was rough, bouncing the Cat and jerking her wing with stabs of turbulence that made me think of the ordeal she'd come through on the water. Blue was thinking the same. He called the cockpit on the intercom, and Harry passed it on.

"Blue wants to know if we can keep her out of the turbulence, after the beating she had on the water."

I thought for a moment. I didn't want to give away that height, after she had used a lot of power to reach it. But it was getting rough. The tops were rising. Might be some cu. nim.* ahead. I had seen a reflected flash of lightning in a gap. Better go down.

"Yes, Harry. We'd better go down."

Jack had some steaming-hot baked beans for me, with peas; and salad. I sat in the swinging chair by the chart table and really enjoyed this meal, while Harry took her down through the cloud.

For a quarter of an hour she bounced her way down through five thousand feet of height and came out of the bottom at fifteen hundred. It was better there. Harry brought up

---

* Cumulonimbus—turbulent storm cloud.

the throttles to cruising power and settled her down close
below the cloud base.

I finished my meal; then went up to relieve Harry so he
could come below and have his share of Jack's creation.

She was cracking along now, bouncing a little, but not encountering anything really rough. Light from the moon was
stabbing through the cloud breaks, casting a silver net to the
floor of the night. I couldn't see the surface clearly, but had
assumed the wind to be southeast and had allowed for drift
accordingly.

A screen of absolute darkness came in towards her, and
soon there were shining particles flying into the light of the
windshield. I took her off the auto-pilot and into a heavy
shower of rain. The rate-of-climb indicator moved up the dial
as the cloud base began to suck her up. I eased down the
power and pressed the nose down to keep her out of the tentacles that sought to draw her into the turbulent cloud. A
glare of lightning flashed on the screen of night, and was
gone as she drove on through the blackness with the sound of
rain on the hull like tearing canvas. It certainly would have
been rough in the tops. She was better below, wading along
in the rain.

Harry came back when he finished his meal and we both
sat up there for a while.

Three hours had passed now since we left the island, and
ahead the night was opening up again. The high cloud was
left behind us, and only scattered, fine-weather cumulus cast
its shadows on the moonlit sea.

Back through the tunnel hatch I took a sight on a flare,
and found five degrees of port drift. The moon was almost
dead ahead then, and the air fairly smooth; so I used it with
the sextant to find our distance run. The result was disappointing, showing a groundspeed, so far, of only ninety-five
knots; but we knew how we were going now, and what drift
to allow. We held on for another hour at fifteen hundred feet
to be sure that the improved conditions were likely to continue; and then climbed her to clear the scattered cloud at
five thousand.

With apparently reliable weather ahead, we settled to our
routine of running the aircraft. Harry on watch in the cock-

pit, Blue and Angus at their stations, and I checking her along with the navigation.

For the first time I began to think of South America as a material destination; as something that could actually happen out of the events of the past two days. As the hours passed through the night I began to tie-in the ends of the navigation to give us an accurate picture of the aircraft's progress. We could hardly miss the continent of South America, but our first objective was the small island of Juan Fernández, which I expected to sight about ten o'clock in the morning, local time.

I kept a check on our distance run, mostly from sights of the moon, which was well situated ahead in the early night; and later, astern, Arcturus, Alpha Centauri, and whatever star was conveniently abeam, gave us a running check on how the aircraft was holding her course, in a night that was perfect for navigation with smooth, clear air over low and scattered cloud, at five thousand feet.

Each time I looked up to the cockpit Harry's figure was silhouetted there against the lighter night outside. How he kept awake I do not know, because there was little action to give him the stimulus to suppress the long-overdue sleep he needed. I could see his hand occasionally reach out to touch the turn-knob on the auto-pilot, and the parallel bars on my master compass showed that he was never off the course I had given him.

We must have been flying through the eastern sector of a big region of high pressure. Our star position showed that the winds, still from south with an easterly component, were gradually decreasing in force as we made distance to the eastward.

As the night went on to the early hours of the morning the cloud thickened below us and we passed over black holes in a floor of light cloud that shone silver in the light of the lowering moon. This silver surface began to rise slightly to the level of the aircraft, so we let her work on up gradually on cruising power, skimming the magic carpet below us on smooth and silent air; for the roar of the motors had long ago passed beyond our conscious hearing and would have been noticed only if it had suddenly ceased as a sound.

An hour before dawn Jack brewed us some coffee, and Blue came down from his watch to pump up the fuel from the hull tanks. Everything was normal on the engineer's panel and there had been no more sign of oil-pressure failure from the compensating valves. I had forgotten about it, actually, in the stress of other events. But in the air Blue lived with the gauges and fuel cocks and all the visible life-system of the aircraft, and he never let up on his vigil, except for a brief descent from his throne in the tower to perform some job in the hull of the aircraft.

I went back to the blister for the last information the heavens could give us before the dawn. The two friends of the night were still there, the moon astern and the brighter pointer to the Southern Cross abeam in the south. I was particularly careful with these sights. This was the last position we could fix to set course for Juan Fernández, and it had to be right. Conditions were perfect, and I went forward to work these sights with a sense of a new beginning. They gave us the position at 1034 G.M.T.: latitude 31° 22′ South, longitude 91° 00′ West. We had run a thousand miles from Easter Island in exactly ten hours, against an average head wind of sixteen knots. Longitude less than a hundred degrees now. That really seemed to be Eastern Pacific. And latitude about the same as Port Macquarie. At Valparaiso it would be almost exactly due east of Sydney. The drumming of the aircraft seemed now to be laying the west-east line, to haul in the main cable of the new airway, encircling the world through Australia in the southern hemisphere.

Jack was watching me lay down the position lines on the chart. He made me feel like a conjuror who has just revealed the result of some magic trick, while he himself was the associate who made the appropriate abracadabra gesture to a spellbound audience.

One of the most critical decisions that ever confronts a navigator is to act upon the evidence of his calculations when it means some drastic last-minute alteration of course to make a small island, after everything has been going smoothly up to that point. Fortunately, it rarely happens; and when some surprising position results from sights and calculations it is immediately suspect and can easily be checked. I would like to be able to give the impression that navigators never make a mistake in calculations; but it so happens that, being human, they do. One of the most accurate measures of a good air navigator is the calm and the speed with which he recognizes and rectifies a mistake when he has made it. Sometimes you have to be quick, calm, and persistently accurate, in moments when it would be easier to throw up the whole thing, and pray.

Now the evidence from the star and moon sights was consistent with the main pattern of previous results. At a former position the aircraft had been set twenty miles south of the track, mainly by a reduction in wind force which could not be observed by the drift sight because of cloud. I had altered course for Juan Fernández from this position, and now this last check from the fading heavens put us on the converging track for the island. It was good magic for Jack, and it put us in the frame of mind to think about South America.

"How are we going?" Jack asked.

"Slow, but we've got plenty of fuel. Looks like we might even arrive about the time that was mentioned in that message at the island. Gives me the horrors now, to think of that boat trying to come alongside."

"That wailing cry from the whaleboat in the darkness. That was the worst."

"Yes. That and the crashing noise in the tail as she came down on the seas. Anyhow, it's behind us now; and we're all set for Juan Fernández. Six hundred and thirty-two miles to go at ten thirty-four z."

"What time are we due at Juan Fernández?"

"Fifteen thirty-nine z on our present groundspeed." I checked off the longitude with the dividers.

"That would be about ten twenty-three in the morning at the island."

The early, white light of dawn was high in the east now. I could do no more with the navigation; so I suggested to Harry that he should take a spell.

Alone in the cockpit I disengaged the auto-pilot and flew the aircraft on her course into the east. There were no breaks in the cloud now and the tops had risen, till at eight thousand feet *Frigate Bird* was skimming the surface with a delicious sensation of speed as the white cloud top slid by closer below her. Ahead, the surface was pink in the rays of the rising sun, and occasionally she would fly through the shining veil of a wisp that rose a few feet out of the main cloud tops. The sky was intensely fine, with the clear, reliable appearance of a perfect world. It was an inspiring beginning to the new day, with the incredible prospect of South America rising with the sun, the darkness of years behind us; and, in the aircraft, a peace that was complete fulfillment and perfection.

This was the use of aircraft I had pictured, flying west from the stricken Rumpler thirty-four years ago. *Frigate Bird* was the envoy of friendship, touched with the glow of the morning light that came out to meet her from South America.

Beyond the cloud horizon I saw in my mind the long, narrow territory of Chile, with the great Andes barrier to the east. That had now been surmounted by the regular internal airlines of South America. But westward there still existed an isolation from the new world of the south some infinite distance across the Pacific, and beyond the daily thoughts of Chilean people.

That moment of time, there in the cockpit of the Catalina, was worth all the resources expended to reach it.

Later, Harry came back refreshed, and took over his watch again. A position line from the sun ahead gave us confirmation of our progress, with the stepped-up groundspeed we had expected.

We were approaching now a position abeam of Más a fuera, a small island ninety miles west of Juan Fernández; and, at this time, conveniently, some breaks in the cloud appeared.

Through one of these, about forty-five degrees off the starboard bow, we saw the unmistakable shadow of this island. The top of the land was not visible, but the line of the shore was definite at the base of the shadow in the blue haze under the cloud. I entered it in the log: "1440z. Más a fuera bearing 140 degrees magnetic; distant 20 miles." It was gone in a few minutes and was obscured before it came abeam.

We were now intrigued to discover Juan Fernández, the home of Robinson Crusoe; to see the fabulous place that inspired Daniel Defoe to write his story of the shipwrecked sailor, Alexander Selkirk. We pictured a place of color, of beaches and palm trees; lush vegetation and brilliant tropical waters.

Before Juan Fernández was due we were searching the gaps ahead for another shadow that would disclose its coast. The cloud was breaking up now, and on the water the drive of the wind was from north. About ten minutes before the E.T.A. we saw the island in the distance ahead. From twenty miles away it looked dark and uninviting, and not the sort of place we had imagined.

As we closed in to come over the rugged coastline, cloud still hung on the 3000-foot peak of Crusoe's island. The hills were bare and inhospitable, and the coast was rough, with rocky bights in the shoreline where huge patches of kelp stained the blue of the ocean.

Frigate Bird came in over the western end of the island, and flew on over the same bare hills and razor-edged crags to the eastern end.

We turned, circled the rugged landscape, and came back to take our departure for Valparaiso from the eastern point of Huesco Ballena. The only signs of life were the white dots of many goats grazing precariously on the steep sides of the mountains, and this was the last we saw of Juan Fernández as we straightened up on the course for Valparaiso, and the white floor of cloud again closed in below us.

To fulfil my original conception of the South Pacific crossing, I had kept Valparaiso as our final destination. Many years ago in the early days of the British Colony in Australia, George Bass, who in a small boat with Matthew Flinders was prominent in exploring and charting the Australian coast, had set out from Australia on a pioneer voyage to link the new land with South America. Bass, with Valparaiso as his objective, disappeared; and as far as I know there is no reliable record of his fate. But I thought that now, in the new age of the air, we couldn't do better than have as our objective on the west coast of South America the seaport for which George Bass had set out from Australia. Only recently we had learned of the Chilean Air Force flying-boat base at Quintero and I had been invited to land and leave our air-

craft there. From the messages passed to us at Easter Island
it was obvious that we were to be received at Quintero. So, to
keep the spirit of the earlier thoughts and impressions, I decided to make our landfall at Valparaiso, circle the town, and
fly up the coast to land at Quintero.

With Juan Fernández and the South Pacific Ocean behind
us, we flew into the last stage with free minds and an easy,
downhill feeling in the aircraft. She was light now, having
burned more than three tons of fuel, and she whistled along
on less than half her available horsepower. The last that we
saw of the wind, on the water out of Juan Fernández, was a
fresh northwest breeze that would boost us along and put the
aircraft in ahead of the time we were expected. So I kept the
power down to reduce the airspeed, and she loafed along
with an easy stride.

All desire to sleep had passed, and we began to tidy things
up in the aircraft. I went back and shaved, using all the fresh
water I wanted for a wash. Then as the sun came abeam
about noon, I used it a last time to check on the course for
Valparaiso.

Though we would almost certainly be expected to arrive in
a dramatically disheveled condition I wanted our arrival to be
in good order, since we had worked that way. So we changed
into our uniforms, which had been protected on their hang-
ers. They were a gray khaki, for the sea, and the genuine
thing, the best wool gabardine, woven in Australia.

With an hour to go, and everything stowed and in order
below, I went up to my seat in the cockpit. As I sat there,
with Harry in the other seat, we might, I suppose, have appeared excited and tense with anticipation; but we had no
words for this situation, and no need for expression. Both of
us, I think, had too much inside us for talk. We sat in silence,
looking ahead, waiting for South America.

The ocean below us, and as far as we could see to the horizon, was entirely covered with a level layer of cloud. It was
so flat, white, and so far into the distance, that it seemed to
go on forever, with no possibility of land ahead, or in fact of
earth at all. This effect was increased by the low tops at
about three thousand feet, distant and ethereal below the aircraft flying at seven thousand.

Soon after I went up to the cockpit the faint outline of a single high cumulus appeared over the cloud horizon. It was as we had seen it in the distance before New Caledonia: the cloud of sun-heated air on mountains. Perhaps the Andes were below the cloud. Still a hundred miles to go to the coast and Valparaiso. Another fifty to the foothills of the Andes. Yes, it could be. Cloud on the Andes, a hundred and fifty miles away in the clear air.

Fascinated, I watched this faint creamy-white outline against the pale blue of the sky. More appeared, just touches on the horizon, fading to nothing, north and south. It suddenly hit me. These *were* the Andes. Not cloud on mountains; but the high, snow-capped peaks of the great range of the South American continent.

I touched Harry across the cockpit, pointed through the perspex ahead over the instrument panel. "South America, Harry. The Andes."

We hadn't anything to say. We just looked across the cockpit then out to the Andes. I called up Blue and told him, turned and signaled to Jack and Angus to come and look. We all stayed there, gazing over the cloud top to the white shapes over the horizon that hardened and became the land of a continent as we flew towards it.

Chapter 18

Chile Welcome

WE SAW the aircraft before the coast was visible. Angus had it over the radio that a Chilean Air Force Cat was coming out to escort us in. We had passed them back a signal giving our E.T.A.: that we were coming in for Valparaiso, and would fly up the coast for Quintero after circling the town.

It was just a speck in the distance, at first. Then they must have seen us, because it quickly grew into an aircraft coming towards us: the high wing of a Cat, with the two dots of the motors and the hull slung below.

As the Chilean Cat approached, the cloud layer ended in a definite line across our front and we saw close ahead of us the sun-hazed coast of South America. Through the haze the white shapes of houses on a hillside were faintly visible.

The escorting aircraft came in on our starboard side. She soared up in a turn, showing the gray hull of the Cat with amphibian wheels tucked in her sides, and lined up to fly alongside us. On her rudder we could see the star of the Chilean Air Force. I turned the control wheel and swayed the

283

wing of *Frigate Bird*, and the Chilean Cat acknowledged our
recognition with the same salute.

Then the outlines of ships and the harbor of Valparaiso
came out of the haze.

Valparaiso—the destination of George Bass when he set
out in a sailing ship from the young colony of New South
Wales to make sea contact with Chile. I wondered whether
he had reached it and been captured and held prisoner in the
mountains, according to rumor; or whether his ship had been
lost in the South Pacific. But this was Valparaiso, here below
us, out of the dream of Australia–South America, as I had
pictured it; with wharves and waterfront buildings and the
masts of ships, all of the sea; a coast to the land where
houses covered the hills. I leaned *Frigate Bird* into a slow
turn over Valparaiso; then straightened her up to fly up the
coast for Quintero.

It was only fifteen miles, so in a few minutes we were over
the Air Force station. The Andes had gone as we descended
into the haze. Then there was the coast of Chile. Now
Quintero Bay was below us; the yellow dots of flying-boat
buoys. The landplane strip, the hangars and buildings; many
people in the area near a jetty by the moorings.

Quintero tower was clearing us in. I left the radio to
Angus because I wanted to concentrate on the approach and
landing. I heard Angus's unhurried voice relayed through my
earphones.

"Quintero tower, Quintero tower; this is Able Sugar Able,
Able Sugar Able. Would you pass alighting instructions,
please?"

I heard the tower come back in English directing us in,
with the accepted international procedure. I had to come
back now, from my own thoughts of air and sea and aircraft,
to be part of a system again. It stamped on my mind the fact
that we had jointed the two ends of the airline system over
the gap where no voice came in from the tower.

We passed in over the bay, saluted the station with a steep
turn over the hangars and the jetty, and I took her away for
approach into the bay from the west. A few flecks of white
showed on the rocks by Punta Liles, the south headland of
the bay though the northerly swell was barely discernible on

the water. This was the approach; along the swell. She slid on down the last slopes of the air; the point went by the wing; she floated in silent suspense along the bay; then the water took her with a smoking sizzle on the keel. She ran, slowed, and came to rest from the air, idling through the water.

In a moment Angus was down in the bow. The hatch opened, and the flags went up in the breeze as she turned to taxi in. Over the bow, the red, white and blue flag of Chile with the silver star; and the Australian flag, with the Royal Air Mail below it.*

I headed her for the yellow and black striped mooring buoy off the jetty, and she went on in with that final, down-from-the-air feeling of the Cat coming in from over the ocean.

Our drogues had gone, with everything else, at Easter Island, and cross-puffs came at her in the mooring area; so I couldn't use power to make an accurate approach and we had to accept a line from the launch at the mooring. It just took the highest shine off this final moment; but that quickly passed as we were overwhelmed by the first flood of welcome to Chile.

A launch was already alongside. I came up through the bow hatch, stood on the seat, and passed a hand of greeting to eager enthusiastic faces. I couldn't make contact yet, with any discernment. There were uniforms and gold-braided caps, light suits in the sun, cameras, an obvious black Homburg hat of London, a felt that looked like Australia; but with me still were the silent gray propellers of the engines, with fresh oil streaks under the cowl, the round blister, and an urgent sense of arrival in the rumble of the idling motor in the launch. It was all one impression, of something ending and beginning in a flood of human emotion.

The launch took us away for the air station. I went up the steps with General Gana. It was the same on the wharf. The dam of welcome burst over us. It was smooth and in the order of an Air Force station; but it was spontaneous, genuine, and I felt here in this first reception the fulfillment of

---

* This time we did carry an air mail: the first official air mail between Australia and South America.

the human purpose for which we had flown from Australia.

We walked on up, by guards of honor, past the hangars for the mess.

I handed over the air mail formally to General Gana, and there was silence; then the songs of our nations, as the Chilean and Australian flags were raised before the standing people. General Gana, who I afterwards discovered flew his own aircraft on his missions as Director of Civil Aeronautics, showed his understanding of our situation by telling me that quarters were ready for us to sleep in as long as we liked; but first there was a party for us in the mess. I had last felt tired when Jack produced coffee before Juan Fernández; but now I was riding on air and ready to join in the reception which awaited us.

The gathering there was in high spirits; quick, individual conversations with many people, who it seemed had long been awaiting this day of contact with Australia. There were men and women of wide interests, converged to a point upon this first aircraft ever to fly in from over the great ocean to the west.

A quiet, significant figure was Cortínez, the first man to fly over the Andes. I felt an immediate contact with Cortínez, who had led the way for the air services which now had surmounted the Andes barrier and ended the isolation from the east. The press interview I gave went out with the news of *Frigate Bird*'s arrival, to spread the whole front page of the main newspapers.

There was full recognition of *Frigate Bird*'s flight. We received, from our own country and from the United Kingdom, many telegrams of congratulation, including messages from His Majesty's Government of the United Kingdom, the Commonwealth of Australia, and the Leader of the Opposition in Australia. We greatly appreciated these messages.

I had no notes for a planned speech. I didn't need any. All that could be said was fresh, in the great welcome we were experiencing. I will not attempt to reproduce what I said, because nothing could recapture the spirit of that day and hour, though it will live with me for ever.

As the freshness of this reception continued through the afternoon, word came through that the President of the Chil-

ean Republic had diverted his private aircraft, in flight for
Santiago, and would be landing at Quintero to welcome us to
Chile.

Down at the airport, we watched the D. C. 3 come in and
soon afterwards met His Excellency the President, Gabriel
Gonzales Videla, and Señora Rosa Markman de Gonzales,
his wife. I showed them over *Frigate Bird,* which now had
been hauled out and stood at the top of the ramp by the hangars. We discussed details of the flight, and its future significance; and I knew at the end of the day that much of our
purpose had been achieved.

After dinner in the mess that evening I met for a few
words with Parragué before going to bed. I think the time
was about eight o'clock.

It was several hours before I went to my room.

I found that Roberto Parragué was a practical idealist. For
years he had planned the flight out to Easter Island and acquired the necessary knowledge and experience to make it.
At last, in January 1951, he had set out, as captain of *Manutara,* for the island. Group Captain Barrientos, officer commanding Quintero, was senior officer in command of the
flight. After a successful flight to the Isla de Pascua, a point
of urgency concerning the return flight to Chile made an attempted departure in adverse conditions necessary, and in
this attempt the aircraft was damaged. Parragué's mind
looked beyond the horizon, and I knew he'd be there again.
We talked of many aspects of the air; came round eventually
to its effect on our domestic lives. Though sometimes there
was anguish in departures, I told him that I had to fly. We
agreed on this point, and when we balanced it all against a
regular, organized life on the ground and he told me of his
home and his wife and four young children, he summed the
whole thing up perfectly when he said: "The routine: it is the
best enemy of love."

It was about midnight, on Monday, March 26. I began to
realize that I had not slept for four days and three nights,
since Thursday night at Mangareva. I reached the bedroom
with sleep really closing down on me now. I opened the window and looked into the darkness over Quintero Bay. Then I
undressed and instantly passed out on the bed.

A series of meetings and functions had been arranged for us in Chile, and the question of our future plans had to be decided.

Before leaving Australia I had discussed with Group Captain (now Sir Thomas) White, Minister for Air, the possibility of continuing the flight, on an international relations basis, across the Andes to Argentina, Brazil, and other South American states. During the last year communications had come back to Australia from various parts of South America showing considerable interest in this first step towards regular air communication across the South Pacific, and it had been agreed that visits to other states, after the arrival of the aircraft in Chile, would have a worthwhile value. The necessary government arrangements had been made to provide for this extended mission of the aircraft, and it was left to me to decide whether to go on or to return from Chile.

Soon after we arrived, I had been asked whether we intended to return by another route, with the obvious inference that we would certainly not attempt another flight by Easter Island. This thought was persistent enough to impress me with the need to return that way, to prove our flight and to make the double crossing of the South Pacific by the regular air route of the future.

During the struggle for survival of the aircraft at the island, I could not imagine the possibility of making another attempt to fly that way if by some miracle we were lucky enough to escape from it once. But now the whole thing seemed more possible, particularly if the radio beacon could be relied upon as an emergency for bad weather, and if we could accelerate the fueling, now I knew that no special equipment had been provided by the oil company, and that we were wholly dependent upon the good efforts of the islanders. At any rate, this doubt which existed about the possibility of success on a return flight by Easter Island convinced me of the need to return that way.

The season was, however, moving towards winter, and the sun, on its seasonal passage into the northern hemisphere, was pulling the westerlies up from the southern ocean. Easter Island, in latitude 27° South, was coming within the region of rapid changes of wind and weather which occur between the tropical trade winds and the westerly winds of the southern ocean. In summer it is known that easterly winds are reliable at the island, but now the unreliable weather was beginning to prevail. If we were to return by Easter Island we should have to sacrifice the pleasing prospect of an extended flight through South America and leave for the return flight within a few days.

The alternative was to go on through South America, and return by the regular North Pacific route, which I regarded as an anticlimax. I thus decided to attempt almost immediately the return crossing by Easter Island. Later, perhaps, another flight could be undertaken, for the international relations purpose, through South America.

A thorough inspection and overhaul of the aircraft was needed and this was made easy for us because the Chilean Air Force offered to undertake the work. This fine gesture by our friends in Chile allowed us to leave *Frigate Bird* entirely to their engineering section at Quintero, and to accept the wonderful hospitality extended to us in Santiago, Valparaiso, and other places.

I decided to fix April 3 as a tentative date for departure. With a full week ahead in Chile, we went on up to Santiago to keep some of the appointments already arranged for us.

General Celedon, Commander-in-Chief of the Chilean Air Force, placed at our disposal, for our personal transport, a Catalina amphibian. With this aircraft we were able to move freely between Quintero and the Santiago airport of Los Cerrillos, where cars also were standing by for our use. From a base of quite incredible comfort and luxury at the Hotel Carrera we moved into the life of Santiago.

At a simple but impressive ceremony, General Celedon conferred upon us the privilege of honorary officers in our several capacities, Harry and I receiving the colorful designation of Piloto Guerra Honorario de la Fuerza de Chile.

I was also presented with a Chilean Air Force officer's cap. Escorted to the cap maker, who worked in the tranquil setting of an old walled garden in the center of Santiago, I discreetly selected a rather good-looking but low-ranking officer's cap; but my escort, a Chilean Air Force officer of authority, immediately said, "No! You must not have that! You must have a General's cap!"

I have worn that cap since 1951, and I shall never wear another as long as I have this significant token from the Chilean Air Force.

During these few days we found ourselves in the modern city of an advanced and cultured civilization. There is a sense of exclusiveness there on the narrow west coast of South America, with the snow peaks of the Andes standing high in the east, and the surface of the Pacific extending to infinity beyond the western horizon. In climate it is the California of South America, and in the air is the same expectancy.

On our third day in Chile, General Gana entertained us to luncheon at the Mabille Casino. This was a representative gathering of many leading personalities in Chilean life. Besides being an inspiring and enjoyable occasion in the beautiful garden setting, it gave me an opportunity to reply to General Gana's welcoming speech, to tell fully of our purpose and our hopes for the future through the regular air services that would eventually follow the flight of *Frigate Bird*.

At this gathering, particularly, I was aware of the desire for contact with Australia. The words that were spoken, from both sides of the South Pacific at that luncheon, were not designed as mere formal expression for diplomatic purposes. They came from the hearts of men, on the one side inspired by the flight of this aircraft to their country, and on the other by the flood of welcome the arrival of that aircraft had released.

In Santiago our visit was just one continuous activity, and I began to feel I would have to get back in the air soon for a rest. Wherever we went—in shops, in the streets, the hotel —there was excitement and happiness.

People grasped us by the hand as we passed, and spoke, in Spanish, half intelligible words of tribute and enthusiasm. A man sweeping the street stopped me outside a store where I had gone to buy some Chilean dolls for my children, and the traffic was held up while he grasped my hand and we sorted out a laughing conversation in English and Spanish. Down in Robinson Crusoe's Tavern, under the Carrera Hotel, there would be some new personality, some new contact with our flight and our visit to Chile; sometimes the press or the radio, sometimes a pilot from the airlines, men and women joining with us in a general celebration.

Then up in the starry enchantment of the roof restaurant, with soft lights on the tables, and the harmony of good food and Chilean wine.

We received invitations from the Palacio de la Moneda to dine with the President and Señora Markman de Gonzales.

When, at a simple ceremony in the palace before dinner, I found, at the investiture by the President, that our services had been assessed in terms of the highest honor that the Republic of Chile can bestow, I found it difficult to express in words just exactly what this meant to me, and, I think, to the others of the crew of *Frigate Bird*.

"Al Merito Bernardo O'Higgins, grade Commendator" is a rare award, of deep significance to the Chilean people. Bernardo O'Higgins became the first President of the Chilean Republic, after he had won for Chile her independence from the rule of Spain. I accepted it on behalf of the crew, who had made it possible for me to guide *Frigate Bird* through a most unreasonable ordeal of endurance, the crew working as a team, with a fixed and cheerful intention to reach our destination whatever the circumstances.

One does not set out with the purpose of achieving distinction. That would be a superfluous load to carry on top of the plain intention to arrive. But it would be nonsense to deny that to receive it is a pleasure. I was overwhelmed by this immediate and high expression of national recognition for our services.

If I allowed the reaction of regret to touch the decision to return so soon for Easter Island, it was based on the increased incentive which our reception in Chile had given to an extended visit in other South American states, and on the possibility that General Gana would be able to accept my invitation to join us in *Frigate Bird* on that cruise. There could be no compromise on this decision, however. Return via Easter Island stood clearly defined as a need, to consolidate the results of the flight and to place it on a foundation that could never be shaken. We could return later and undertake an extended flight through South America.

Santiago, Chile, was obviously destined to be an important trans–South Pacific airport of South America. This evening with General Gana we explored the effects of the South Pacific link, and I think we both saw the tremendous possibilities it opened up for human contact.

The following morning the crew of *Frigate Bird* went out of Los Cerrillos for the last time, in the amphibian piloted by Lieutenant Núñez of the Chilean Air Force.

Before leaving Los Cerrillos I had discussed with Señor Rene Méndez, the Chief Meteorologist, the trend of the weather between Quintero and Easter Island. Though the reports were, as he knew, limited to the Isla de Pascua, Juan Fernández, and South American coastal stations and ships, a fair idea of the main weather pattern seemed to be possible for the route. The extensive high-pressure system which we had experienced seemed to be a dominant condition for the time of the year, and it was likely that we could ride round the top of a high and carry favorable winds almost through to the island. Against this, we should probably encounter higher cumulus tops this way than through the center, where the winds would practically balance out to nil in their effect.

This was the prospect when we left Los Cerrillos for Quintero, where Señor Enrique Torrealba, forecaster for the Chilean Air Force, would give us the final forecast and synoptic chart.

The work on *Frigate Bird* had progressed almost to the point of a test flight. The aircraft had been given a complete overhaul during the week, and the engineers at Quintero had worked continuously to have her ready on time. I had acquired from the navy store at Valparaiso some new heavy chain and shackles, and two new Northill anchors at Quintero. Blue had changed the oil pressure compensating valves for new ones, and Angus had checked over all his radio gear.

Through the efforts of the Australian Embassy in Washington, a set of four JATO rockets had been obtained from Panagra. These rockets had been flown in from Lima, and we now had them attached on the hull for the westbound takeoff at Easter Island.

Everything went perfectly on the test flight, and we came

in and left her on the head of the ramp ready for departure.

Then the inevitable happened. No airplane can leave on this sort of flight without something coming up at the last moment. Jack had let it drop, casually, that there had been some doubts at the station concerning the condition of the wing attachment bolts; rumors that they had been stretched by the plunging of the wingtip floats in the sea at Easter Island.

It had been reported to Harry that when the engineers had come to the nuts on these bolts in the course of their routine check, some of the nuts had come up about three turns for the correct tension. I knew there were only two things that could cause this. Either the bolts had stretched, or the wing had compressed. The bolts held each wing onto the center section, and stretched bolts are likely to break under tension. Immediately, I saw delay: unstable weather at Easter Island, everything snow-balling to a drawn-out departure from Chile instead of a clean, scheduled start of the motors, and take-off. By a miracle of luck, we had come in only a few minutes off schedule. I wanted to depart in the same state of order.

The chief engineer didn't happen to be about at the time, or he could have told me. Where was the man who personally tightened the nuts? Could I see him, with an interpreter? Yes. That would be easy.

He was there in a few minutes.

"Can you please tell me the greatest number of turns you made on any nut?"

Answer, through the interpreter: "One."

"How many nuts did you tighten as much as one turn?"

A little thought. Then a clear answer: "Three."

"Did many other nuts tighten nearly as much as one turn?"

"No."

"How much, then?"

"Perhaps some nuts a quarter of a turn."

A quarter is well inside the tolerance. The situation passed.

The nearest we could go to expressing our thanks for everything was to borrow the Quintero Casino mess to give a party for our friends there. I presented the significant Australian pilot's wings to Group Captain Barrientos and to Commandante Roberto Parragué.

That night after dinner at Parragué's home, he took from the mantelpiece over the fireplace a carved wooden figure of the *manutara* bird. It had been given to him by the people of Easter Island, and he now passed it to me with good luck for our return flight. This was a personal gesture which touched me very deeply. I decided to mount the figure above the instrument panel of *Frigate Bird II,* where I already had a small silver plate with the name inscribed, "Norman Birks, Frigate Bird." Norman Birks of *Frigate Bird I,* who had been with me on the Clipperton Island flight, had died very prematurely soon after the war, but his name flew with us in *Frigate Bird II.*

On the norning of April 4, a favorable route forecast was given by the meteorologist, and fine weather was reported from Easter Island. There was some doubt about how long the Easter Island weather would last, but in the absence of any reports from the west it seemed unlikely that we would gain anything by waiting another day. So I decided to take off at three o'clock in the afternoon, with the intention of making an early-morning landfall at the island, a quick refueling, and departure the same day for Mangareva. A second overnight run should then put us in the region of Mangareva late in the night, or early in the morning.

The station commander entertained us with a farewell party at noon, and then by the time we had collected our personal gear and the many gifts we had received, it was time to go down to the aircraft, which had been launched and now was waiting for us on the mooring.

Chapter 19

Easter Island Again

FRIGATE BIRD was low in the water, with the rockets on her sides and all tanks full. She was going to be heavy out of Quintero. I wanted every gallon of fuel she could carry, in the event of bad visibility at the Easter Island end.

I watched the long lines of the northerly swell moving in and breaking on the beach. In the western corner under the headland it was calm, but with the heavy overload we needed distance to run for take-off, and most of the run would be in the open bay.

A plan was forming in my mind as we went out in the launch. I had been thinking of this take-off and watching the habits of the sea in the bay while we had been at Quintero. To attempt a heavy-load take-off straight out to sea with the advantage of a clear getaway would obviously be a heaving, bouncing disaster without rockets; and we needed the rockets at Easter Island. It would have to be along the swells, with floats up immediately there was air control. Quintero Bay is really only a bight in the coastline, about three miles between the headlands, and completely open to the sea except for the southwest corner, where the seaplane moorings are. A lightly laden aircraft could use this corner, with a short run for take-off, but *Frigate Bird* would need most of the bay for her run on the water, and clearance over the base at the end. There was a light wind from west in the bay. It looked like a take-off crosswind on a long curve off the beach, to keep her running along the swells.

We were alongside then, and going in through the blister. Everything had been stowed and I had set up the chart and prepared for flight when she had been on the ramp at the base. We went straight to our stations and settled in to work the aircraft.

I sat for a few moments in the pilot's seat and let my eyes move around the cockpit; from the faces of the instruments, by the auto-pilot controls, the throttles and propeller controls in the roof, the radio panel with the intercom switch, the handles and pointers of the trim controls. I took my piece of clean rag from the niche under the cockpit window and wiped the wheel till it was receptive to my touch. I turned the controls through their full travel and felt the freedom of the rudder with my feet.

Over the bow, Angus had the flags out: the blue peter for departure and the flags of Chile and Australia. Harry was there, and Jack standing, as he did, looking out through the opening in the bulkhead behind my head. We passed a "this is it" look, and I called up Blue. He was ready with the engines. I gave him "contact port" and heard the whirr of the energizing starter. The blades began to turn over, and the engine came in.

The aircraft was alive again. I signaled to Angus and she moved away from the buoy, turning.

"Contact starboard" to Blue, and the other engine came in, picking her up to head away down for the far end of Quintero Bay.

I took her round the curve of the beach, just outside the rising swell before the surf, to test the surface there for take-off back over the base. *Frigate Bird* slid on, awake and eager, pushing the sea ahead of her half-sunken nose.

As we passed out of the sheltered corner she rose and sank down the swells, which fortunately were long and low and only rising as they reached the shallow water off the beach. But they were fanning out into the bay in a series of curved undulations parallel with the curve of the beach.

"I'm going up to the northern end of the bay, Harry, for take-off along the swells towards the base. We'll get Blue to whip up the floats as soon as there's air control, so I can have the wingtips free for a long turn on the take-off run, following the swells round the beach. We're in the swell already, so we might as well go up near the end of the beach and take all the run we can get."

Harry nodded his understanding, looking into the distance up the beach. "Will you want to go above maximum power for take-off?"

"I don't think so, Harry. I want to save the engines from that if we can. We'll take all the available distance and see how she goes. If it isn't looking good, I'll give you the signal and you can let her have the full treatment."

I turned to Jack, with one of those last-minute thoughts. "Tunnel hatch closed, Jack?"

He gave a definite thumbs-up signal. I'll bet he had checked it; looked at it; checked it again; gone out through the bulkhead; and gone back and checked it again.

I turned her in the northeast corner of the bay, and started into the take-off run.

She plunged into the sea, showing no inclination to get up and go. Heavy water came over the nose, so I could not see the beach. It was normal reaction with such a heavily laden aircraft, but we were close to the surf of the curving beach, for a run to pass inside the headland. Steering on the gyro heading and thinking of the surf off the port wing, I went with her, trying to coax and lift her into the nose-up attitude before the planing run. The wingtip floats would have to go up. I snicked the switch for the signal to Blue as the water began to clear away and she started to go. I had her now, riding high and seeing the way. Then the nose rode over the bow wave; and down. She was away, really singing over the surface, the engines roaring with wild and far-flying sound as they knew they were in it now with flight the only escape from the terrific effort of take-off. As the floats tucked into the wingtips I laid her on a steady turn to follow the swells

round the beach. It was wild, terrific, screaming sound with
the starboard wing down and close to the surface like a seabird skimming the ocean.

I watched the base coming closer; and the air speed rising
on the dial. The seas were passing under her, trying to break
the rhythm of that last terrific effort that builds the few extra
knots for flight. I followed the surface with the bottom of her
hull, to keep it running cleanly for those knots. The base.
The speed. Base. Speed. A light tension on the control column, testing out her readiness to go. She had the speed now,
but the surface was holding her. It was time to break that
grip, and she was asking to go. I drew back the column and
lifted her away. She sagged and I went with her, easing and
letting her flatten for speed, just above the surface. As soon
as she was really flying Harry drew off the power to maximum climb; then I held her down till the needle showed a
convincing speed on the dial; and lifted her over the hangars
of Quintero base.

Everything asked for straight and steady flight till the load
had eased, but we had to salute Quintero. The air was
smooth, so I laid her on a turn round the 300-foot hill of
Centinela and came in low over the base. Faces and hangars
and places we knew looked up to the soaring Catalina, and a
B.25 was taking off on the strip. I turned her westward over
the bay and straightened her up for the ocean.

The air was gray in the west, with many thin, indefinite
layers of stratus cloud and a low overcast on the sea. We
topped the white floor of this fog cloud at a thousand feet
and leveled off with reduced power to ease the engines after
the effort of take-off. She stole on into the gray air, back in
her endless stride already; beginning to eat into the two thousand miles for Easter Island.

Suddenly a shape came into the corner of my vision, and I
saw a B.25 come up alongside. There was another off to starboard, and the faithful Cat amphibian soaring westward with
us. I had flown the B.25 with R.A.F. Transport Command
during the war, and could see the struggle the pilots were
having to keep down to the speed of the Catalina. The fine
wing of this aircraft was leaning on the air, trying to hang on
at the little more than a hundred knots we were making.

Our escorts stayed with us till there was danger of collision when thin films of cloud began to close in around us. Then they soared away on the air, and were gone. *Frigate Bird* was alone again, in the gray mists of space.

I went below and made the first entries in the log.

        2009   Cast off mooring.
        2023   Airborne.
        2026   On course.

On course about half past three. About an eighteen-hour flight. Difference in time about two and a half hours. Should be in the region of the island at seven o'clock in the morning.

This time we were not concerned with Juan Fernández. We were on a track of 280 degrees True, direct for Easter Island. Through an occasional break in the cloud, the southerly wind we had expected was visible on the surface of the ocean. The breaks were not big enough for an accurate drift sight, so I made an estimation of six degrees' drift to starboard and allowed for this on the compass.

We flew on into the darkening air with little sight of surface or sky, climbing slowly to four thousand feet, to fly in a clear stratum between the cloud layers.

Four hours out from Quintero the first stars showed for a few minutes secretly through the fine gauze of a high overcast. I saw a bright star through a gap where Procyon should be, and I took this for the first check on the aircraft's track. It showed us to be ten miles north of the estimated position, but I let her go on without altering course because I believed that the southerly would be easing up towards the west and that later sights would show us to be picking up the leeway.

I decided, however, to keep a continuous check with star positions whenever there was visibility of the heavens. The weather looked unreliable, and I wondered how far we might have to fly for the island from the last sight of the stars.

There was the radio compass; and the most emphatic instructions had been passed from Valparaiso to the Easter Island station for beacon operation extending over a period from four hours before our E.T.A. until the aircraft had landed. But I remained unconvinced that we could rely upon the limited radio facilities at Easter Island to bring us in. The station there had not been installed for airways use, but it was able to transmit on the international distress frequency of 500 kilocycles, which was within the reception frequencies of our radio compass. There seemed to be no physical reason why it should not fulfil its normal function in guiding the aircraft in over at least the last two hundred miles; so I looked forward to using this aid if we needed it; but the strategy of navigation had to be based on the assumption that it didn't exist.

It was a night of anticipation: watching for breaks in the cloud above or below. Picking off stars as they came; watching; slipping aft to whip up the tunnel hatch quickly to snatch a drift sight; feeling again that sense of control of the situation, then having it pass away in cloud again; wondering about the winds. Did she still need that drift allowance, or was she making some unknown leeway, setting her far from the track for the island? There were some known facts; some unknowns; some uncertainties which had to be accepted. Harry sat for hour after hour holding the course that I gave him, sometimes varying the altitude of flight to give me stars, sometimes diverting to avoid the dark shadows of high and incalculable cumulus tops in the moonless night, but always keeping a record of time and course on the diversions so that I could plot them with the estimated groundspeed to calculate where we had finished up.

Blue was there, eternally on watch over the sensitive life-system of the aircraft: a sentry on whose alertness the life of the aircraft and everybody in her depended at any moment of the eighteen hours of flight.

And Angus, with his mysterious contact, touched a world which for me had gone with the Cat and the B.25's as they swept away in the mist; he sat there in a trancelike state, sending out our position to a listening operator where men in a tower were plotting our progress towards the Isla de Pascua.

Jack brought up the coffee.

"How far out are we?" he asked with a quizzical smile. I took the pencil and touched a point on the chart, stepping off the distance on the latitude scale.

"About halfway, Jack. A thousand miles. A little out of gliding distance to the land."

Blue was sending up the fuel from the hull tanks now, and Jack went back to relieve him on the pump so he would not be out of touch with the engineer's station.

And so we went on through the night, till in the early-morning hours before the dawn the air cleared and the brilliant heavens opened up before us with stars right down to the horizon ahead.

We were flying now at seven thousand feet, fourteen hours out from South America, and only the shadows of fine, scattered cumulus passed in the night below us. The aircraft, after the long hours of almost continuous movement through solid and broken cloud, was now set in the night in absolute stillness.

The star positions were checking along like a navigator's dream, each one dead on track and giving a constant and convincing groundspeed of 118 knots. In a gap before the night opened out I had taken the planet Saturn, and the star Alpha Centauri; and these confirmed the series of other fixes.

An hour and a half later, with only an hour before dawn, I sighted the sextant to stars for the most critical calculation of the flight: for the result on which we must rely for a final course for the island. All the stars were there, so I accepted the guiding light of four that would give a good cut for the lines of position.

As I returned to the chart table I felt well satisfied that they would give us an accurate fix; and when I had worked them, the four lines intersected almost at a point; in latitude 27° 58′ South, longitude 104° 23′ West. This position was eight miles south of the track and 260 miles from the island.

I handed Jack, for his records, a chit with this position and the E.T.A. Easter Island as 1325z—about six o'clock at the island.

With the very last of the night I picked off Vega to the north, and the line confirmed the converging track for the island. That was the end of it. Now it was the compass and the steering.

But ahead the whole aspect of the weather was changing.

There was a high overcast, and mountains of cloud hung with black and menacing bases apparently flat upon the ocean.

At 1230z I estimated that we were 138 miles from the island. There appeared in the early light of dawn to be conditions of nil visibility ahead. I began to feel now that we would really like to use the radio compass. Invisible winds in the cloud could drift us to run close by the island without seeing it through the rain. The needle of the compass would home on the Pascua station and guide us in.

I called across to Harry, who was looking ahead assessing the conditions, and he obviously wasn't pleased.

"We'll tune in the compass to the Pascua station now, Harry, ready for the run in."

It was the worst time and conditions for the radio compass; dawn and heavy electrical cloud with rain. As I tuned the equipment in, there was a convincing flicker of the needle. It began to move round the dial, hovering for a moment over the heading of the aircraft. We were all watching; Harry, Angus, Jack, and I. It swayed around zero as though seeking something it knew was there. Then suddenly it swept round the dial and stopped, quivering on 180 degrees.

Easter Island dead behind us!

We'd passed the island.

We all looked at the needle, intent; then not believing anything for an instant. Nobody said anything. We just went on looking at the radio compass, which emphatically said that the whole of my navigation was wrong; that there must be some bad, consistent mistake somewhere that had been putting us far behind our true position.

Possibilities flashed through my mind.

The chronometer.

No. Not that. Angus had been receiving and passing time signals to me; and anyhow the daily rate of four seconds a day was constant.

The tables.

It wouldn't be the tables. I'd checked the month and date every time I looked out data from the air almanac. No, there couldn't be any consistent error there; or in the facts from the H.O. 214 tables which I'd been using for years.

Then I thought of the sextant: whether it could have been

dropped and was recording an error in the star altitudes because of some damage in the mechanism or mirror adjustment. But it couldn't be that, because I knew it hadn't been dropped. Nothing had happened that could damage the sextant.

No. It was none of these things.

The radio compass was wrong. Its indications would have to be ignored. I looked back to the needle; still showing the island astern; still telling the story that for every minute of flight on my course we were leaving Easter Island farther behind in the clear morning and flying into a sea of rain and cloud ahead. The whole thing seemed crazy. But there was no compromise. We would have to go on; just hold the course till the time the island was due; and if it wasn't there, then . . .

I could see Harry. His face was expressionless, but I knew what he was thinking. I think he believed in the star navigation, but here was weather in which the final approach of any aircraft to its destination would normally be on the path of some radio aid, to which we were both accustomed on the airlines. Now the only available aid was flatly contradicting the other means of navigation and the aircraft was thrusting on into sightless air, in defiance of that aid.

"We'll have to forget the radio compass, Harry. We'll go down, and check her through on the drift sight. There's no reason to alter course." My own words seemed hollow to me.

Very soon afterwards the needle began to wander again. It never showed the slightest stability or gave a definite indication again, so I switched it off to avoid its distracting influence and we let *Frigate Bird* go on down through the height for the sea.

The altimeter showed three hundred feet before we could see under the cloud base, which hung in a series of tumultuous rain storms in jet-black curtains to the ocean. Between were lighter patches where the surface extended for two or three miles in the gray drizzle of rain.

The wind had swung round to a blow from northwest, and I immediately laid off twelve degrees for southerly drift at a guess from the streaks and the white caps on the surface. Then I went back and confirmed it with the drift sight, and we just sat there and held on the magnetic compass a course for an island that was supposed to be behind us.

There was half an hour to go now till our E.T.A.; still about sixty miles. Right on our track ahead was an enormous cumulo-nimbus cloud with its black belly almost on the sea. Everything ahead was obliterated by this monster, and in the blue-black rain on its front, columns of ocean were being sucked up into its base in waterspouts. It was a grim and threatening spectacle. I took over the flying and leaned *Frigate Bird* into a turn to avoid this cloud.

We flew for five minutes on a diversion of forty-five degrees to starboard, passing, in moderate rain, the northern edge of the cloud base. Round behind it, the same conditions existed as ahead, forbidding a turn back to cancel out the diversion. The whole flight was closing in too quickly now for involved, two-way operations between navigator and pilot; so I made a quick guess allowance for the diversion and straightened up on what I estimated would be a converging course for the island. Ahead on this course were only moderate showers, with visibility varying from three miles to a few hundred yards in the rain. No diversion was needed for this, so we just flew on the new course, wading through the showers and running out again to regions of visibility.

I looked at my watch. The minute hand was past the hour. 1308 G.M.T.

E.T.A. at 1325.

Seventeen minutes to go. It was closing in. Suppose we were early. The showers were heavy enough to obliterate the island. Drift from the last star position had been hard to check. We could run by it; not know it was there. The temptation to wander round every shower to keep visibility was almost irresistible. But that was confusion. It would have to be resisted. We must hold dead on the course now through anything; and wait for the E.T.A.

Always my eyes went back to the hands of my wristwatch. It was twenty-one minutes past now. Only four minutes to go. The sea was swept by a wind of thirty knots from north. I had fifteen degrees on for drift. The Cat was crabbing across the ocean. There was nothing but sea, and wind, and rain, and the aircraft. Land in this region of infinite loneliness could not be real; just a dot on the chart as a theoretical objective. The existence of an island seemed like pure fantasy.

The time came; and passed—1328 on the watch.

The minute hand moved now towards nothing: for widening spaces of empty ocean and air; but I knew that the strong wind change would make us a few minutes late on our original E.T.A for the island, though everything around us said it could not be there.

We ran into a tearing shower, the stream shrieking at the screen and the hull. There was nothing but sound and emptiness where water enclosed the aircraft in some endless ocean of rain.

Then it lightened to a film of fine drops cast in a net of sunlight, and immediately before us was the solid outline of land.

Harry was the first to see it. He pointed ahead to the island standing between sea and cloud. There was no doubt about it. Here was the sudden, incredible miracle of the Isla de Pascua, clear in a moment above a live, blue ocean where low and broken cloud marked the passage to fine weather in the west.

The sudden change held us silent for a few minutes. Now everything was the island, where before there had been nothing.

We came up to the low cliffs on the southeast side, and swept on over the grassy hills. The end of the weather hung over the island in cloud so low that it rested on the higher hills. We threaded our way across by the valleys and flew out over the ocean on the western side. The scene was the same as on the eastbound flight. A northwest wind had roughened the swells on the side where the fuel lay at Hanga Piko to a surface which was too rough for landing.

Ovahe Cove again.

All the madness of the last night there came back to me. The wind and the banging of the tail; the eerie cry from the whaleboat; the roar of the bombora where the seas broke near the aircraft. I could feel the anchor lines snapping one by one as the seas hit her in the wild darkness with the moan of the wind and echoing boom of the surf on the cliffs behind the tail.

Now it was quiet again, luring us in under the cliffs, waiting to catch us again with a wind change, helpless on the anchor and with *Frigate Bird*'s tail to the rocks. Thoughts flashed into my mind of trying to reach Mangareva without landing for fuel; but they were immediately canceled by the simple facts of distance and remaining fuel. The seas of the island stared up to us, knowing that we had to land for fuel.

I turned her up the west coast, flying for the North Cape and round by Anakena Cove, where there is the only beach on the island.

It was all forbidding and hostile to the aircraft, and as we came round again over the southeast coast it became obvious that there under those same cliffs of Ovahe Cove was the only place where we could land safely and be within an hour's whaleboat passage of the fuel.

But before landing I wanted to confirm our previous observations of the island contours and general features, for the landing strips that would have to be constructed for aircraft refueling on the Australia–South America airway. So for the best part of an hour we circled the island, crossed its land where we could fly below the cloud base, photographed its significant features, and sketched in added information on the large plan we had brought out from Chile.

With this survey completed I took *Frigate Bird* over to circle Hanga Piko, to confirm our arrival and immediate need of fuel. Then we flew back over the island, passed above the cliffs of Ovahe Cove and out to sea. The landing run was obvious, with westerly puffs coming out of the bight in the coast. I turned her in with the floats down and slid down the air for the calm area under the shelter of the cliffs. The surface was easy, with the long swell coming across our line of approach. I kept her flying low above the water till the cliffs were coming in. Then I drew off the power and let her down.

She surged to rest off the rocks of the boat haven and turned with the engines passing back their sound to the air. The hull was heaving in the ocean. The surge washed lazily over the rocks and the blue water was brilliant with shafts of sunlight striking down to the sinister boulders and caverns below. As it had been before, it was calm and beautiful and warm in the sunshine; but, I didn't like it.

"Harry," I said, "we've got to be ruthless here today. There's only one objective: to get the fuel into the tanks and take off. Nothing else matters. We'll keep the pressure on till the hosepipe is in the tank and somebody's pumping fuel. I just hope those boats will be round to us soon with the drums."

We anchored on the same sand patch. It was half past seven in the morning.

I cannot describe the frustration of that day on the water at Easter Island. We lay there for eight hours before the boats arrived with the fuel. Then it was only the fine efforts of the islanders, in loading the drums in their small boats and making the rough passage round from the exposed side, that enabled us to refuel before darkness again came over the island.

About noon I saw the change coming, from the south. The westerly had faded out, and a light scud was flying over the island at about a thousand feet, coming in from southwest. Soon it had thickened to a complete overcast, and the wind had swung into south.

The sea in Ovahe Cove was awake now, and working up again to imprison the aircraft in a broken turmoil with her tail to the cliffs. I stood on the wing, torn with frustration, watching for the dot of a fuel boat rounding the southern point of the island.

I thought of all the possibilities. Start up the engines and taxi round to the west side. No. By the time that side was sheltered enough to refuel there it would be too late to leave. Take off out of Ovahe Cove and land off Hanga Piko. No good; for the same reason. By the time the Hanga Piko side was smooth enough for landing it would be too rough at Ovahe to get off without rockets; and we needed the rockets for the heavy-load take-off.

No. There was nothing to do but wait and hope that the fuel would arrive in time.

I went down and lay in the darkness of the tunnel compartment to see if I could store up some sleep for another night's navigation to Mangareva.

I must have slept for about an hour when I heard somebody call, "The fuel boat's coming."

Out in the blister compartment again I found the sea shrouded in a fine rain and the cloud base down to the hills of the island. The wind was still in the south, freshening now and coming in squalls, with the sea working up and the whole picture a critical one for take-off even if we could refuel before darkness. A boat was coming up from the south cape of the island and would reach us in about a quarter of an hour.

A second boat appeared. That was better. All the fuel would be there. Close in by the cliffs to the south was still a stretch of sea that seemed possible for take-off. If we could send the fuel up quickly we might get away with it.

It was close to four o'clock when the boats came up to the aircraft. Rain was coming in driving showers and already the threat of approaching night was creeping up on Ovahe Cove. There was too much sea to have them alongside; so they anchored wide off our bow and let down on the anchor line towards us till the hosepipe would reach the tanks.

The boats were things of wet and steaming men, with water and fuel drums, heaving and rising in the seas.

I stood on the bow hatch pressing the willing, smiling islanders to start pumping. Everybody was out, saturated and streaming water; doing something to rig the pump, open the drums, keep off the boats, and somehow to get the hosepipe from drums to aircraft, and somebody pumping fuel to the tanks. The wind came down on us with blinding showers and cloud down to the cliff tops. Everything eastward was the bleakness of sea and westward the dark walls of cliffs ending in the cloud.

We kept at it; plunging the suction pipe from empty drum to the hole of a full one where they knocked off the filler cap, shielding the gasoline from the rain with their bodies. Blue and Harry were on the wing with the big filter, passing the fuel to the tanks. Islanders pumped till they were exhausted; then changed places with others to the pump. Nearly a thousand gallons of fuel had to go up to the tanks of *Frigate Bird*, to give us a good margin for emergency the other end.

It was all a sweating, steaming effort of men and pumps and boats, surging and swaying in the seaway, grasping at boat and aircraft to protect the thin shell from collision with the heavy whaleboats; never letting up an instant to get those gallons aboard.

And they finished it in an hour, by five o'clock.

It was a matter of minutes now, with the wind and weather rising to storm conditions. The instant the last of the fuel was in I called to the crews, and they swung in with quick realization of the need to clear away. We hauled in the hosepipe and drew everything back into the hull. The engines were turned over, and we left tracks of water as we went quickly to our stations.

Now it was a drive for take-off. I felt up and going with the aircraft, exhilarated and confident that she would beat the seas and the rain and the island: that the power and the rockets would blast her into the air and she would stay there. Visibility was almost nil, with cloud down to a hundred feet from the sea, and rain below it.

Blue came in with the engines and we broke out the anchor and bore away for the warm-up. I could just see to the base of the cliffs. I felt the power of freedom in the throttles; contact with the defiant motors as we checked the switches and the props. We had the bearing of the island point and the gyro lined up to the compass for the run down the cliffs.

"All set, Harry. We're going."

I kept the throttles for the first surge of power, then passed them to Harry and used both hands on the wheel. She plowed forward, buried her nose in the ocean, and the propellers went wild with vigor and sound. She had to come up out of this. I pressed the button for the first rockets and she leapt up out of the sea and started to drive on the surface. A sea was coming. She had to go. The second rockets. She hit it, bounced; there was smoothness under her, and nothing ahead. My eyes stuck fast to the flight instruments. Something was happening to the aircraft; I felt pressure on the rudder going through my foot and mind to the gyro heading. Full left rudder and she was still slewing right for the cliffs.

"Trim, Harry! Rudder trim!" I shouted, holding her over with both hands and full left rudder. I could just stop the gyro turning as she blasted on, blind, on instruments, out of the sea, and madly trying to turn and ram the cliffs.

It was all too quick for reasoning. I had to stop her turn-

ing. Just that; and hold the airspeed that would stop her
dropping back in the sea. The violent action on the rudder
and wheel hard over to stop some blind madness as she projected herself into the air was the only reaction. Nothing was
visible outside but rain. The sudden need for flight on the instruments came as a flash with the slew of the aircraft. Then
I seemed to be holding her with full left rudder through my
foot and eye and the gyro heading. I thought of the rudder
trim, but could not let go the wheel for a second to check it.

Then she must have passed the end of the invisible island;
she seemed to be coming back under control. Still in cloud, at
about three hundred feet, I got her settled down, in straight
and level flight on the instruments. Harry had reduced the
take-off power and she showed a safe margin on the airspeed
indicator. The rate-of-climb indicator was settled on zero.
The heading was steady, and we might have been on a normal instrument flight at eight thousand feet instead of roaring
along a few hundred feet above the ocean.

The rudder trim was neutral, and the aircraft was behaving
quite normally again, but bouncing in the turbulence of
squally rain and cloud still on the southerly heading. I turned
her slowly round to pick up the heading of the compass
course westward; and as she straightened up we flew out into
clear air below a black and threatening cloud base five hundred feet above the ocean.

Chapter 20

Westward Home

EASTER ISLAND had gone. The lonely Isla de Pascua had passed from our sight in the take-off, and now was invisible in a shroud of swirling wind and rain somewhere only a few miles behind us.

"The rockets, Harry. They must have failed on the starboard side, to cause that terrific slew."

"I didn't see them; but I saw you ram on the rudder immediately we left the water." Harry and I laughed across the cockpit. We both felt the same sense of relief that *Frigate Bird* was still in the air.

I called up Blue.

He hadn't seen them either. But he had now jettisoned the rockets; so we had lost nearly a thousand pounds of weight already. That situation was behind us, but the day ended in the west in an apparently impenetrable wall of darkness where the black front of violent weather shut right down to the sea. Behind it somewhere must have been the red brilliance of sunset; but here, against the damp, gray daylight, it was blacker than night.

All idea of holding the course for Mangareva was wiped
out by this front that lay across the track. We could have
plunged straight into the copper black of the swirling cloud
base, which hung like the take-off weather at Easter Island,
with little clearance above the sea. But the aircraft was still
heavily overladen; she had received another inevitable beating
in the take-off; and ahead was the violent turbulence of wild,
unstable air that would snatch at her overloaded wings and
threaten to tear them from the hull. Escape from this weather
was the first objective.

"No future in this, Harry. I'm going to divert." And I
turned her north to fly along the front.

As we closed with the lowering cloud base, black claws of
wind darted out of the darkness and snatched at the lead-gray
ocean as though seeking some victim from its surface. Advancing squalls shook the wings of *Frigate Bird* and flicked
the tips with jerking bumps that warned of the main attack. I
had to pull her more away to keep stable air, and soon we
were heading northeast and losing distance from the west.

Something caught my attention on the port wingtip. There
was movement in the retracted float each time the wing was
jagged by the turbulent air as we sought to approach the
front for a way through towards the west. The float was retracted, but it hadn't locked up, and with every bump its
downward jerk shook the wing with a flicker of vibration.

She was in the air, flying; seeking a way to get back on a
course for Mangareva. I tried to put the float out of my
mind, but every time she hit a bump it was there in the corner of my vision, jerking the wing. The whole thing had
passed far beyond normal human reaction. The frustration of
waiting for fuel with the sea working up to pin us down at
Ovahe Cove, where I knew the aircraft would not survive the
approaching night. The wild abandon of take-off. The mad
slew for the black wall of cliffs. A breakout into the clear
with the rising hopes of escape and a good departure for
Mangareva. Then the black barrier, shutting us off and turning us first north, then northeast, losing ground, away from
the track to our objective. Now the jerking float, with visions
of a moonless night and turbulent cloud that would threaten
to tear the wing off the airplane. I was down to fundamen-

tals. Nothing mattered but the present, just to steer the airplane clear of the front till a gap would let us through.

Even the basic navigation had to be scrapped. The turning
and weaving were too indefinite to trace back accurately
from the times and headings. I just kept a thought of the general direction we were making and let it go at that.

About forty miles along the front, a gap appeared. It was
only a lighter shade in the black curtain and there was nothing but rain behind it; but it was an entry to what seemed
like ordinary weather with rain. I turned her for this break
and she bore into the stream of rain.

At least it was action in the right direction with the gyro
heading west on 270 degrees. In a few minutes we broke out
into the clear again. Ahead was another black wall flat down
to the ocean, but we could see round its ends, where the
lighter rain was colored faintly pink, from the distant red of
sunset. There were signs that we were breaking through the
barrier; that perhaps there was fine weather to the westward.

Frigate Bird nosed her way low over the ocean, and passed
round the end of the darkness ahead. She flew on through a
shining sea of lightened rain, and out over a clear surface
where a steady wind from south swept the ocean clean.

Ahead the horizon was clear with the last red light from
the sun that was now well below the plain of sea before us.
Great castles of cumulus cloud stood high over the horizon
against the colored sky, predicting a broken night, smooth
with stars, then the sudden blows of turbulent air with closeness of swirling cloud smoking over the exhaust flames.
Harry and I looked ahead into the west.

"Looks like we'd better go up now, Harry. If you take her
on up I'll give you a course in a few minutes."

As I went below I felt *Frigate Bird* respond to the high
rhythm of climbing power, and through the perspex panel
above the navigation table the darkening surface of the ocean
began to sink away as she ate her way into the height. I
switched on the spotlight over the table, looked at the chart
where the dot of Easter Island lay, and began to piece together the ends of the beginning.

I gave the course to Harry.

We were beginning to smooth out the ends, but the wingtip

float kept coming back at me, and the cumulus towers ahead were fading from our vision into the shadow of approaching night. When Harry leveled off at seven thousand feet the distant tops were high above our level, and I didn't see how we could avoid a bouncing when we met them. I went back to the cockpit and called Blue on the intercom.

"Blue, I still don't like that port float. It moves every time we hit a bump and it just doesn't look as though it's locked up, to me."

"Yes, Skipper, I can see it jerk from here; but it's fully retracted."

We'd discussed it before. That seemed to settle it. Then I had an idea.

"Is there any reason, Blue, why we can't wind it hard up with the hand gear and lash the handle so it's held in the up position?"

It was a makeshift thought, but it might work.

"I'll try it, Skipper."

I waited while Blue strained it up with the handle, and lashed it fast with some rope. I watched out under the wing and waited till we passed through some slight turbulence in the gaps between the tops of cloud. The float didn't move now. It stayed streamlined into the wingtip and *Frigate Bird* seemed tight and going again.

The stars were coming, but the air was rough as we sneaked along among the high shadows of cloud. When we emerged from the maze of diversion through the weather, I had been content to know that we were heading westward, making distance for the region of Mangareva. But time was passing now. The brilliant light of Sirius was dead ahead of us, and Canopus crept out from behind a cloud top away in the southwest. I was able to reach these stars from the cockpit, and managed to average out some acceptable sights in the doubtful air conditions.

The results surprised me. We were not north of the direct track as everything had suggested; but twenty miles south, and not so far on as I had expected. We were nearly four hours out from Easter Island and had not been able to obtain any drift sights since climbing to seven thousand feet. Either the position was wrong, or the wind at seven thousand was

from the opposite direction to the southerly we had seen on
the water. Again, it was A or B. One has to believe something; so I believed the star position.

A quick check showed that unless the aircraft encountered
strong head winds we would be over Mangareva before
dawn. So far, we had no weather from the Gambier Islands,
so I had to provide for a daylight approach to allow for
weather which might be bad for a night let-down for these
islands.

To arrive in daylight we should have to hold at some position, or shuttle on the track at some stage of the flight. I
thought of the little atoll at Oeno as a possible holding point.
Though it is only two miles wide, the light patch of the shallow lagoon had been very clearly visible in the moonlight
going out, whereas Henderson and Ducie islands had been
less distinct. Now, in this moonless night, I thought we would
have a better chance of seeing Oeno than any other island,
and it was in a good position as a holding point for approach
to Mangareva.

So, from our calculated position in latitude 26° 22′ South,
longitude 115° 10′ West, I set course for Oeno, 850 miles to
the westward, and 220 miles from Mangareva.

In the back of my mind also was Pitcairn Island, which lay
sixty miles south of the track for Oeno. To fly over Pitcairn
would be a point of interest in the flight, but it could only be
kept as a possibility.

This was a night of intense concentration for Harry. Only
for brief intervals were the cloud tops below our level. Then
he was able to let her go on the auto-pilot, holding my course
for Oeno. But for most of the night he was hand-flying the
aircraft, peering intently into the coal-black darkness for the
tops of high and turbulent clouds.

Sometimes the brilliant night would be studded with stars
to the horizon. Then an intense darkness would creep up
slowly into the heavens and wipe them from our sight. He
would strain and look into this obscurity, seeking the edges of
great anvil clouds, where he could turn *Frigate Bird* to fly
through the canyons between them. Lightning would suddenly flash to us a clear-cut scene of these huge mountains of
the air; and in that second our minds would photograph the

enormous buildups where we sought the valleys and crevasses to avoid the violent turbulence of cascading air inside the cloud.

Harry never let up for a moment; and after each wandering, seeking passage through these mountains, each involving perhaps an hour or half an hour of diversion from the course, I would go back to the stars again and find our position; and again we would set off on a new course for the little world of Oeno, somewhere in the hidden depths of this strange and awe-inspiring night.

I stood by Angus as he was calling Mangareva: calling, calling into that eerie world of the radio where no call had ever gone out before. I wanted Mangareva weather, as a basis for a final decision about Oeno: whether we should hold for daylight, or not bother about Oeno and go on through for a night approach and landing. If Mangareva happened to be clear, we could locate over the main island group in the darkness, then find Aukena and the strip of water along the beach, which would be faintly visible even without a moon. The training in blind landings in the Saint Lawrence River, proved in the return to Gander Lake, had left me with confidence for night landings without flares. We could drop a drift-sight flare to mark each end of the alighting path clear of the coral, complete the circuit quickly while the flares were still burning; then go in on an instrument landing.

That part of it was clear before us. But in cloud and weather, without the audible road of a radio range, or even a radio beacon, I wouldn't want to be mixed up with those island mountains until daylight.

Angus was calling Rikitea: calling, tuning. Calling; tuning. Calling, with the urgent, dramatic staccato of Morse from the limitless spaces of air in the night, for a man with earphones in a room at the settlement of Rikitea who could tell us whether there were stars over Mangareva.

I watched Angus, and waited for the information on which I would make the decision. His hand was moving; writing those quick rushes of signals from his ear to the radio log.

He had Rikitea. I could decide now. He handed me the signal; and I read, "Mangareva weather. Wind calm; sea smooth. Lightning and rainstorms in area. High cloud ⅜ cumulonimbus. Height of tops unknown. 6/8 cumulus; base 2000. Tops unknown."

The decision was obvious.

Oeno and a daylight approach to Mangareva.

Or Pitcairn? We'd have to alter course now if it was Pitcairn; with two hundred miles to Oeno, a hundred and fifty to Pitcairn.

I went up front to the cockpit, closed the bulkhead door, and drew the night curtains close over the cracks where light comes through from the navigation compartment.

As my eyes adjusted themselves to the darkness, I could see that the air ahead on our course was fairly clear. There were a few dim shadows of isolated high clouds which could be avoided with slight diversions. But away in the southwest, in the direction of Pitcairn, were continuous flashes of lightning that illuminated the sky on a front of more than a hundred miles. Each glaring flicker of the lightning lit up the hard edges of high storm clouds rising to levels many thousands of feet above the ceiling of the Catalina.

From the moment of entry to that maelstrom there would be no possibility of fixing the aircraft's position; and drift observations would be impossible. We would just have to crack right on and hope for the sheer luck of sighting Pitcairn, which in the moonless night would not be visible unless we hit it off dead on the nose. And if we didn't see it, departure would have to be made for Mangareva from an unknown position, to an island where we knew there would be a high percentage of nil visibility.

Oeno was different.

It was small but we had the means of finding this island. Luck was not the deciding factor. We had stars, and a pilot who would steer my course, exactly.

Pitcairn was out. It would have to be Oeno.

I turned to Harry, and lined up the situation. "Pitcairn's out, Harry. We'd really get a beating in that stuff down there on the track; and no way to find the island. Angus has Mangareva; but it's not so good. Cu-nim with rainstorms, and a lot of cloud. Best we hold on for Oeno, locate, and wait over the island, then go on for a dawn approach to Mangareva."

Harry nodded; reached out his hand to the turn-knob of the auto-pilot and checked her to the compass course.

"What's the E.T.A. Oeno?" he asked.

"Twelve thirty-one Greenwich, Harry. About three forty-five local time. Sunrise at six zero eight. If we hold for half an hour over the island, that should give us a good daylight approach to Mangareva."

The heavens were clear to Oeno. The brilliant pointer was there by the Southern Cross; Alpha Centauri was dead abeam. I altered course three degrees from this last star line, for Oeno; and waited.

Harry and I were silent, up in the darkened cockpit. We turned off the fluorescent instrument lighting, cut out any sign of glare that would lessen our night vision; for everything was now building up to arrival over Oeno, and I had no illusions about the difficulty of sighting it without a moon. A single drifting cloud below us could cover it from our sight at the time of passing. It could hide in the night a few miles off the track of the aircraft, or it could come in and pass under the nose and away without being seen.

I kept a secret anticipation about this tiny atoll. It was not only a visible point on which to use up time without wandering about over the ocean, but it was a rabbit I had to pull out of the hat. I believed we'd see it unless it was under cloud, but the limits were fine and if it happened to pass three miles away I might as well have missed it by fifty.

Inevitably, as one does when expecting land to appear, I was looking for it ten minutes before it was due; and unreasonably, as the E.T.A. approached, I began to feel it had passed. There is no strong, silent act about this navigation. The whole fascination about it is one's vulnerability to the emotion of expectation. There'd be no fun in it if one felt infallible. The thrill of every landfall is really the exciting confirmation of a belief, and certainly not evidence of one's infallibility.

It was too dark to see the ocean. Night was an all-enveloping region of darkness, punctured by occasional stars from above; but I thought of that white sand of the lagoon and the break of surf on the encircling reef, and felt they were signs we might be able to see from seven thousand feet.

I was back on the watch hands again. At 1230 I was straining and peering into the darkness, feeling it might be slipping away; 1231 passed and there was still only the black void below and around us.

The second hand was sweeping the dial at 1232. I didn't want to look across at Harry, but I wondered whether he could possibly have missed it, from his side.

I looked on out into the distance. The second hand was turning inevitably, confirming the passage of another minute. I opened the sliding window at the side and peered down in the darkness, shielding my eyes from the roaring airstream. Directly below I saw a lighter patch in the darkness which for a moment I thought was low cloud.

But it was Oeno. The circle of surf was definite, enclosing the lighter patch of the lagoon; and within a small stain of darkness that must have been the wooded island in the lagoon. Scattered clouds were around and below us in lighter patches in the night. They appeared as dim islands of the air; but this was an island of earth: ethereal, but fixed on the floor of darkness below us.

I took over from Harry and turned her up on a wingtip, looking down to the strange outline of this island of the night. We lost Oeno for a few minutes as a cloud drifted over it on the light wind that must have been blowing from south. But we kept on circling and picked up the island again when the cloud had passed.

Half an hour later we left this passing anchorage of the air, and set course for Aukena and the light of dawn. *Frigate Bird* flew on with the high towers of cloud, passing through rough air in some of the tops, but she was lighter now and better able to take the shocks of the swirling air.

As the dawn came up behind us I decided to descend and make a low approach for the Gambier Islands, which appeared still to be under the heavy cloud reported by Rikitea radio. We could not afford to be contemptuous in our approach to this larger objective, even from only 220 miles, because it too could sneak away under the rain clouds and stay silent and invisible as we passed.

Frigate Bird stepped lightly down from seven thousand, through broken cloud; and at the bottom we came out over a shining, windless ocean, quiet and empty to the distant horizon. We took her on down to five hundred feet.

The ocean shone like a giant disk of polished shell, and it moved with slow, awakening impulse from the depths. There was infinite peace and a quiet satisfaction in this dawn. We flew on the knowledge that the great hazard was behind us, that *Frigate Bird* had confirmed her eastward passage with the westbound flight, which now was approaching the calm of a sheltered lagoon and the warm white sands of Aukena.

Jack had cooked us some bacon and eggs, and we seemed again to be humans with normal reactions instead of impersonal sources of energy.

We passed through light showers of rain, and out again to distant views of the horizon, where these gray screens hung to the glistening surface from the base of broken cloud.

Somewhere beyond this early light was another Pacific mountain, with the peaks of Mangareva standing above the surface; but here everything was submerged and invisible beneath the deep polish of an apparently endless ocean. *Frigate Bird* flew steadily on her destined compass heading, tracing a thin straight line across this infinite scene. In the night there was no suggestion of earth, but an intimate sense of security in the aircraft committed to flight in space. The aircraft stole with silent breath on the light of dawn into this vast and lonely region.

I was waiting for sight of Timoe Atoll, which was due close abeam before Mangareva. We were committed to our compass course since the stars had gone; so we waited and watched into the distance, for land behind the rain.

Soon after six o'clock I saw a light streak on the ocean where no rain was falling. It was off the port bow and seemed about fifteen miles away. As we approached, the white light turned to yellow, and a line appeared against the sea. It was the unmistakable ring of an atoll island, though its low land was not yet visible.

"Timoe, Harry." I pointed; and we both seemed to wake from some endless rhythm of night, and cloud, and dawn, and the drumming of the aircraft.

Blue had been there all night, staring at the faces of his instruments, watching the fuel consumptions, descending for a few minutes to normalize himself with movement; then returning to his watch again. He had looked down, with us, to the ethereal shadow of Oeno. Timoe would be something for him, too.

"Engineer from Captain," I called, to attract his attention; and his voice came back, "Captain from Engineer. Go ahead."

"Blue, look out from your port side and you'll see an island in a few minutes—Timoe. It's a little atoll about twenty-five miles before Mangareva."

Angus and Jack came up, and saw the gem of Timoe as it came in abeam. It passed quickly out under the port wing, a little paradise of brilliant blue and yellow and green, unaware of the world. We couldn't see Mangareva yet, but it had to be there; behind the rain ahead.

Frigate Bird drove into the shower. The flying water sang like a hurricane in the palms. She drove on through it, sightless in the rain. Then the stream began to dissolve; a dim shadow appeared ahead. The flying drops passed; and clear ahead in the morning light was the prehistoric backbone of Mangareva.

She was over the reef in a few minutes, and Aukena was there. I took her on round the north end of the island. It was still and lovely by the beach. We straightened up and flew on, to make a turn over Rikitea for those who were expecting *Frigate Bird.* Then I leaned her away, across the lagoon. She flew back, over and round Aukena. I slid her in close by the point, drew off the last of the power, and she floated quietly down on the glassy surface.

We taxied in over fabulous caverns and cliffs of brilliant coral, for the light green of sand off the beach.

Angus was ready with the anchor, and fifty yards off the beach I called to Blue, "Stop engines."

"Anchor, Angus."

She drifted on slowly, drew up the slack on the anchor line, and came to rest.

This was not exactly the end of the flight; but it was the end of the darkness and exposure to the ocean where no flying boat was meant to be. It was the end of the silent region of the air where nothing seemed to exist but the aircraft, where everything depended upon the finest margins, and on vision of the stars. Mangareva, the outpost far beyond the previous flight of aircraft on the outward route, was now the refuge at which we had returned to relatively normal flight.

It was nearly seven o'clock in the morning. The early sunlight was touching the beach over Aukena. A whole day and a night were ahead of us before the flight for Tahiti. I took off my clothes, dived into the crystal water of the lagoon, and swam ashore to the sunlit beach.

The luxury of this day was quite extraordinary, with the aircraft lying peacefully on her anchor off the beach. Passing showers wet our bodies, but in a few minutes we were dry again in the sun. I walked the beach to the north end of the island and felt the cool touch of sand on my bare feet. The sun and the breeze on my body completed a perfect exhilaration in which there was no need for sleep; and ahead was the luxury of night in the cool silence of my bunk under the blister, the rain, and the stars.

Early in the afternoon some outrigger canoes arrived, from Rikitea. With the aid of these small craft and their friendly owners we were able to improvise the refueling of *Frigate Bird* from the drums which had been put down for us under the coconut palms by the French Government schooner *Tamara*, from Tahiti.

At the end of the day the canoes left us, and *Frigate Bird* lay alone in a calm peace. We saw again the sunset over the rugged peaks of Mangareva, and the heavens light up with a million brilliant stars. There was no sound but the faint, echoing swish of the scend on the beach of Aukena. It seemed that even the ocean was sleeping; breathing gently in perfect rest. And that was the last I knew till the light of dawn came up behind the close shadow of the island.

This was the day of Tahiti. Before us was a flight by many islands, in the daylight; never more than an hour from the calm water of a Tuamotu lagoon; the aircraft light, and therefore with effective single-engined performance.

I shaved and had an early morning swim from the aircraft. Then fruit of Aukena, with bread from Chile, and coffee from New Caledonia. In the cockpit the controls came lightly to my touch and *Frigate Bird* was eager to go. She slid quietly away from the curve of the beach, and floated again over the blue deeps where coral crags reached up into sight from invisible regions below.

We went through the formal check of the engines and I turned her by the point where we had come in. The propellers took her with their flood of power that lifted the lightened aircraft and swept her smoothly for the southern point of Aukena. The beach and the coconut palms streamed by the wing and a white goat stood on a rocky crag of the island hill. Then I lifted her from the water and Aukena passed.

We circled the lovely island once; then headed westward, climbing, over Rikitea. The great range of Mangareva passed close below us, and we soared out into space above the western reef.

Then it was the ocean again, by a new track up the main group of the Tuamotus. The weather was clear and stable; and we topped the light and scattered cloud at six thousand. *Frigate Bird* settled in to steady flight at this cruising level, passing from island to island as she stepped her way up the Tuamotus.

On his first contact with Papeete radio out of Mangareva, Angus had passed our E.T.A. of three o'clock. I planned to confirm the reliability of *Frigate Bird* by arriving exactly on time at Papeete; so we used our surplus time allowance for a close inspection of Mehitea.

We swept down on the island and *Frigate Bird* seemed conscious of her release from the tension of her long and canny flight through the lonely regions behind her. She responded with sweeping, easy movements, like a fighter. I took her down to sea level and swept her round the island with the wingtip streaking past the rocks. Seabirds rose and flew in tumbling panic from the roaring monster that had suddenly invaded their island.

She slid round the south end of Mehitea and I hauled her up in a stalling turn and poured her down the other side. A schooner was anchored there, and we flashed by her stern almost flicking the wave tops with the keel, but it was time to be on course for Papeete, so we climbed her to five thousand feet and rode along in cool, invigorating air.

Soon after leaving Mehitea we saw the high peaks of Tahiti showing through the clouds.

The great adventure was over.

But it seemed that here was a new beginning, starting homeward from the future meeting place of all the South Pacific airlines.

As the shadow of Taiarapu Peninsula came in towards us I eased down the power for the timing of our E.T.A., and let *Frigate Bird* start her descent for the welcoming shore of Tahiti. We came in with the land at a thousand feet where Mahaena Pass breaks the reef on the corner of the main island. I pressed her on down till she streaked along the shore at a hundred feet, with five minutes to go for the last ten miles. There was no reserve about this. It had to be dramatic. *Frigate Bird* had to come out of nowhere, suddenly, at exactly three o'clock, and roar into sight over Papeete Harbor. We'd given the day from South America, and the time from Mangareva; and she had to keep them.

There was another, rather personal reason why I wanted to be exactly on time in arriving at Tahiti, and for this we have to go back for a moment to the tragic circumstances at Montreal, to which even the wonderful skill and kindness of Dr. Peirce, the radiologist, could not prevent the ultimate and inevitable end six years later. I was, however, eventually lucky enough to marry again and at the time of the South America flight, the lady who is now my wife, had gone out to Tahiti to handle some important affairs of the flight. I had given her an E.T.A. on the flight out from Australia, and much to the surprise of people at Tahiti, her confident belief that the aircraft would arrive at this time, on this day, turned out to be right. Now I had given her an exact day and time for our return, and wanted her confidence again to be upheld at this timeless lotus island, where such an event was regarded as quite improbable.

The minutes were closing as we passed the light of Venus Point. Coral shore and coconuts, the peace of palm-thatched homes, hills, and great green mountains rising into cloud: all passed by the wingtip as *Frigate Bird* flew the last miles for Papeete.

I could see Pirae coming with a minute to go. She was
right. I held her down flat to the water behind the palms.

At Papeete nothing was coming and the time was three
o'clock. I drew her up over the palm tops and she swept into
sight. Papeete Harbor was below us, with the white masts of
schooners and a crowd of people on the quay. I turned her
up on a wingtip and let her do her stuff; just the one steep
circuit of Papeete, and away for the approach from Les Tro-
piques.

I watched for the cross swell, drew off the power from a
steep turn in, and straightened her up as she swept in for the
landing. *Frigate Bird* couldn't go wrong today. She feathered
onto the surface and rode on up for the mooring. I idled the
motors to give Angus time for the bow hatch and the flags.
She moved in the last few yards with the flags of France and
Australia fluttering in the breeze. The drogues slowed her.
The buoy was by the nose. The propellers fell over the last
compressions as Angus snapped the mooring line over the
bollard.

I had counted on a few days relaxation at Tahiti, but it
didn't work out that way. There was great enthusiasm for the
possibilities *Frigate Bird*'s flight had opened up for French
Oceania. Tahiti, already destined to be a focal airport for air
services from Australia and New Zealand, from Honolulu,
Los Angeles, and from Mexico, suddenly was floodlit as the
principal Southeast Pacific airport for services to the great
continent of South America.

Every hour in Tahiti was full, until five days after our arrival there from South America the time came for us to fly
westward again.

On the evening of April 12, a month after we had left Sydney, we gave a farewell party to His Excellency the Governor
and Madame Petitbon, and other friends of French Oceania.
I had a great sense of fulfillment there that night, and a
knowledge that, though the physical part of our venture in its
exploratory form was mostly behind us, an important part of
the human contact was still ahead. My dream had turned to
reality. Here in Tahiti people had joined with us and our
Australian aircraft through the common interest and purpose

of the contact we had made with South America. I wanted to
carry this on still westward, to Australia.

The following morning *Frigate Bird* left the mooring at
Papeete and taxied down the harbor with the blue peter
flying below the flags of France and Australia. On her side
she carried the indelible imprint of the Chilean and Australian flags, their staffs crossed in union of the two countries. I
turned the aircraft down by the western channel in the reef,
and opened up for take-off. She swept up Papeete Harbor and
was soon away.

In a long slow turn back over Papeete we saw the last of
the dream island of the South Seas. It was remote already, in
the setting where I have always pictured it: green mountains
in cloud, sloping down to coconut palms and the sea; the little trading center of Papeete touching the harbor with the
white masts and hulls of schooners, but somehow revealing
the formality of white uniforms and roses of France, which is
its other character.

It was gone in a few minutes, and the sharp peaks of Mo-
orea began to come in towards us as *Frigate Bird* made her
way into the west. We flew on the trade wind, in clear air
above scattered cloud, to within two hundred miles of Aitutaki. Then a dark leaden overcast faced us to the westward.

When we came to the island, the windless lagoon was the
polished base to a dark dome of threatening weather. I had
thought of going on to Palmerston Island for the night; but
decided to stay at Aitutaki where there was good holding
ground in fine sand close to the beach of Ratuatakura Island.

Though the weather was threatening, the evening was perfect and the anchor was well fast in the coral sand. The aircraft was sheltered by the coconut palms from a blow off the
land; a soft beach would be behind her in an onshore wind. It
was a good position, in which I had peace of mind.

We went ashore and I climbed for coconuts, and later
some people of the island came up in a canoe with gifts of
fruit. They stayed on the beach, singing softly through the
night.

From Aitutaki we flew for Satapuala Bay in Western
Samoa, diverting, on the way, to Palmerston Island on a

search for a schooner which had been unable to locate the island in the prevailing bad weather.

There was certainly weather on the Palmerston Island track, but we saw no sign of the schooner. We were relieved ourselves to sight the atoll of Palmerston, which was hiding behind a black squall and came up suddenly ahead of us in brilliant sunshine.

The flight from Palmerston to Samoa was one of those dream experiences where an aircraft flies in a region of blue space below a cloudless sky and over a calm and windless ocean. There is an infinite intangibility about the dimensionless scene and only the aircraft has any material expression. Yet here again, as we so often know it in the nights above the oceans, is the calm and certain knowledge of the eternal source of life of which we are part.

From this tranquil passage over the ocean we started the descent for Western Samoa, and soon were passing up the north coast of Upolu. Huge cumulus clouds capped the mountains of the island, but the air was clear on the coast towards Apia and the house of Vailima which now was our objective.

The southeast wind, descending in broken gusts from over the mountains, bounced *Frigae Bird* and shook her wing as we flew in over the home of Robert Louis Stevenson, now the residence of the High Commissioner for Western Samoa. From a tight circuit over Vailima I straightened up the aircraft and headed for the open water of Satapuala Bay. In a few minutes she swept in by the reef and the silence was broken by the sound of waves on the hull.

We were entertained that night by the High Commissioner and Mrs. Powles; a happy evening, in tune with the purpose of our flight.

But there was weather about, heavy thunderstorms and squalls in the area. I thought of the aircraft on the mooring; of the cable, the shackles, and the invisible chain under the buoy. I could feel *Frigate Bird* snatching at that mooring as the squalls hit her and she rose on the high-water swells in the exposed lagoon at Satapuala Bay. The comfort of Aggie Grey's hotel at Apia had to be put behind us. It had to be another night in the clanging of the hull. We stayed awake with the aircraft through the eerie howling of the rain squalls, but in the early hours of the morning the storms had gone through; the familiar stars were there in the heavens again, and *Frigate Bird* lay still and quiet on the lagoon. We all had a few hours' sleep before dawn came up with another day over the Pacific.

The sun was over the horizon when we slipped the mooring, and ten minutes later took off to head down the track for the Fiji Islands. The aircraft was closing in with the regular North Pacific air route now, and on course for the beach north of Suva where, seventeen years before, the fuel-laden Altair had taken off from the wet sand on her twenty-five-hour flight for Honolulu.

This was an easy 600-mile run, but it so happened that we encountered some of the worst weather on the whole of the 16,000-mile flight to South America and back. *Frigate Bird* plowed through the bouncing darkness of an aggressive front for more than a hundred miles before the weather opened out to show us again the crater-lake island of Niuafou.

Two hours later the eastern islands of the Fiji Group began to show their brilliant patterns on the ocean, and through a gap in the broken trade-wind cloud we saw the beautiful island of Vanua Mbalavu.

Taveuni passed, away off the starboard wing; more colored reefs and islands seven thousand feet below us; and then, as we descended for the main island of Viti Levu, the white sweep of Naselai Beach defined the edge of the low green lands north of Suva.

We came in low over Naselai and with a slow circuit of the beach saluted Charles Kingsford Smith, whose memory was very fresh to me as I looked down at the spot where, in the Altair, we had started into the take-off run for Honolulu.

From Suva we set off for an overnight stay at Lord Howe Island, with plans to go in to Sydney the following day. In contact with Sir Thomas White, the Minister for Air, through signals at Tahiti, I had agreed to fix our E.T.A. Sydney at 3 P.M. on Saturday, April 21, a day and a time when people who were interested in the flight would probably have the best opportunity to see the return of the aircraft.

We left Suva on the morning of the 20th, and set course for Lord Howe Island; but six hours out from Suva Angus

passed me Lord Howe weather, which he had just received
on the radio. It didn't look good. Strong and squally southwest winds, increasing in force, with rising seas. I saw the
shallow lagoon, fully exposed, with weather developing; *Frigate Bird* lying precariously overnight. Unlike the commitment
on the open ocean at Easter Island, this was an avoidable
risk. We had fuel for Sydney, but I knew that a reception
had been planned for us the following day; so we had to lie up
somewhere till the forenoon of that day. There were quiet
lakes and rivers on the coast, but now we were back in the
world of Air Traffic Control and I did not want to embarrass
anybody by using our present free-roving status to ignore the
rules.

So, it had to be Brisbane; where all the formal setup existed for a night approach and landing in the best airline
style. I pinpointed our estimated position on the chart, and
Frigate Bird turned patiently away from her destined heading, accepting our strange habits of wandering about the
ocean.

Cruising in clear air at eight thousand feet, we flew westward for Australia. *Frigate Bird*'s job was done. She had only
to fly in a few hundred miles for the sunset, to rest upon the
Brisbane River.

But in the aircraft there was a sense of expectant emotion.

In our minds we saw Australia again, out of a situation
where our land had been merely the final destination in a
venture. Now, in the clear air ahead, we could almost see the
Macpherson Range, though it were still several hundred miles
beyond our physical vision. We sat, still and silent, and
watched into the west as the sun sank into the red haze on
the horizon.

"At the going down of the sun—"

Always the words and the thoughts came back in the air.

Then, an hour before Brisbane was due, we saw in the last
light the purple outline of Mount Warning.

There was something miraculous about it; something revealing to us the reason for our flight.

The Macpherson Range came up, in clear outline against
the distant sky; then faded slowly with the passing light. I reduced power for the descent and a hush came over the air-

craft as we seemed almost to be gliding in, with the downhill rhythm of ending flight. When the coast had faded we saw a tiny flash prick the dim horizon at Cape Moreton. Instantly there was a personal contact with the land.

Then the spread of lights from Southport and through to Brisbane crept into the scene ahead, and soon the shadow of Moreton Island was below us. The approach quickened now, with all the brilliance of the Brisbane lights; and coming out to Lyton Reach, the dark outline of the river. I took her in round the land airport at Eagle Farm and called up the tower for clearance to land in the Brisbane River. I could hear Eagle tower calling the control launch in Hamilton Reach.

"Hamilton. Is the area clear?"

"Eagle tower; this is Hamilton launch. The area is clear."

"Able Sugar Able; you are cleared to land."

I felt the floats jerk down as we came round by the red city lights near the approach into Hamilton Reach. The flare path was there stretching into the darkening distance from under the hill where the lights of houses and streets shone on the lazy, glistening surface of the river.

Frigate Bird sank in by the light-studded hill. I eased her out of the descent as the first flare came in, and she floated easily down to the river.

We taxied on to pick up the buoy by the ramp at Barrier Reef Airways.

Everything stopped.

Then there were launches, voices, and people.

The unknown also was gone, and we were swept up in a strange and warmhearted simplicity. In it we passed from the aircraft. People we knew were in the launch. Voices. Lights. The sound of slick water and hurrying of the launch. I was talking; answering questions and welcomes, through some voice from my body that was not my own.

It all moved on, in, and through the terminal building. People around us, all welcoming us home.

We were out of Brisbane in the forenoon for Sydney, estimating arrival for 3 P.M.

The second hand was coming up to the minute at fifty-nine as we entered the circuit area at Rose Bay.

Frigate Bird had kept her appointment with Tommy White. But as I looked down at the flying-boat base on the downwind leg I forgot about the watch. The shore was lined with people, and in the base area thousands of welcoming faces looked up to follow the aircraft in, just as it had been at Quintero eight thousand miles away on the western shore of South America. I saw that the aircraft had joined these people, and the track of her flight still lay across the ocean.

Rose Bay tower cleared us in.

The westerly was whistling in over the harbor at thirty knots, rising in turbulent gusts up the hill of Rose Bay heights.

I held her down in a close approach through the rough air over the houses. Vicious light-green squalls panicked over the surface from round Shark Island. I held her securely in my hands with a little power—felt her touch—waited.

The broken water rattled on her bottom like a shower of gravel. But she held; ran smooth and straight on the surface. I slowly drew off the last of the power and let her settle on the water. The blast of the upwind engine took her away in the turn for the base, and she held her head high, striding in for the buoy with the unforgettable swaying gait of the Catalina in from the ocean air.

Ashore at the Base the great wave of welcome swept over us. It seemed that nothing separated the human scene at Quintero from that which we were now experiencing at Sydney. First to greet us at the dock was the Prime Minister, and of course our sponsor, Sir Thomas White, the Minister for Air. Mr. Menzies' presence was, I thought, significant, as that of President Gonzales Videla was at Quintero. Each placed a seal upon the spontaneous enthusiasm of the welcoming people for this first, visible sign of personal communication across the South Pacific.

How long it would be before the airlines would establish regular services I did not know, but we had made the first gesture and it had been received with enthusiasm and good will.

It was thirty-five years since I had shot down the Rumpler

in flames over Brielen; and had looked then across the
oceans. The South American flight looked like the last of the
pattern I saw that day; the last pioneer flight to be made as a
lead to a great international air route; for soon the performance and range of aircraft would reduce to the simplest dimensions any flight across our world. When I stopped the engines of *Frigate Bird* on the mooring at Rose Bay that day I
knew that I was signing off at the end of a great period of
exploratory flights. But I was already planning a flying boat
cruise service in the South Pacific.

Index

commander, 11; fight with Rumpler, 12–15; leads to resolve to use air for peace, 16–17

World War II: selection of bases in Indian Ocean, 120; with

Atlantic Transport Group, 126–148

Wyndham, 104

Yancey, Captain Lon (navigator on *Guba*), 108–120

THE AVIATOR'S BOOKSHELF

THE CLASSICS OF FLYING

The books that aviators, test pilots, and astronauts feel tell the most about the skills that launched mankind on the adventure of flight. These books bridge man's amazing progress, from the Wright brothers to the first moonwalk.

☐ **THE WRIGHT BROTHERS by Fred C. Kelly** (23962-7 • $2.95)
Their inventive genius was enhanced by their ability to learn how to fly their machines.

☐ **THE FLYING NORTH by Jean Potter** (23946-5 • $2.95)
The Alaskan bush pilots flew in impossible weather, frequently landing on sandbars or improvised landing strips, flying the early planes in largely uninhabited and unexplored land.

☐ **THE SKY BEYOND by Sir Gordon Taylor** (23949-X • $2.95)
Transcontinental flight required new machines piloted by skilled navigators who could pinpoint tiny islands in the vast Pacific—before there were radio beacons and directional flying aids.

☐ **THE WORLD ALOFT by Guy Murchie** (23947-3 • $2.95)
The book recognized as *The Sea Round Us* for the vaster domain—the Air. Mr. Murchie, a flyer, draws from history, mythology and many sciences. The sky is an ocean, filled with currents and wildlife of its own. A tribute to, and a celebration of, the flyers' environment.

☐ **CARRYING THE FIRE by Michael Collins** (23948-1 • $3.50)
"The best written book yet by any of the astronauts."—*Time Magazine*. Collins, the Gemini 10 and Apollo 11 astronaut, gives us a picture of the joys of flight and the close-in details of the first manned moon landing.

☐ **THE LONELY SKY by William Bridgeman with Jacqueline Hazard** (23950-3 • $3.50)
The test pilot who flew the fastest and the highest. The excitement of going where no one has ever flown before by a pilot whose careful study and preparation was matched by his courage.

Read all of the books in THE AVIATOR'S BOOKSHELF

Prices and availability subject to change without notice

Buy them at your bookstore or use this handy coupon for ordering:

Bantam Books, Inc., Dept. WW4, 414 East Golf Road, Des Plaines, Ill. 60016

Please send me the books I have checked above. I am enclosing $_____ (please add $1.25 to cover postage and handling). Send check or money order —no cash or C.O.D.'s please.

Mr/Mrs/Miss _____

Address_____

City_____ State/Zip_____

WW4—11/83
Please allow four to six weeks for delivery. This offer expires 5/84.